THE COMMODITY FUTURES TRADING COMMISSION

THE COMMODITY FUTURES TRADING COMMISSION

JASON H. BURNS
EDITOR

Nova Science Publishers, Inc.
New York

Copyright © 2008 by Nova Science Publishers, Inc.

All rights reserved. No part of this book may be reproduced, stored in a retrieval system or transmitted in any form or by any means: electronic, electrostatic, magnetic, tape, mechanical photocopying, recording or otherwise without the written permission of the Publisher.

For permission to use material from this book please contact us:
Telephone 631-231-7269; Fax 631-231-8175
Web Site: http://www.novapublishers.com

NOTICE TO THE READER

The Publisher has taken reasonable care in the preparation of this book, but makes no expressed or implied warranty of any kind and assumes no responsibility for any errors or omissions. No liability is assumed for incidental or consequential damages in connection with or arising out of information contained in this book. The Publisher shall not be liable for any special, consequential, or exemplary damages resulting, in whole or in part, from the readers' use of, or reliance upon, this material.

Independent verification should be sought for any data, advice or recommendations contained in this book. In addition, no responsibility is assumed by the publisher for any injury and/or damage to persons or property arising from any methods, products, instructions, ideas or otherwise contained in this publication.

This publication is designed to provide accurate and authoritative information with regard to the subject matter covered herein. It is sold with the clear understanding that the Publisher is not engaged in rendering legal or any other professional services. If legal or any other expert assistance is required, the services of a competent person should be sought. FROM A DECLARATION OF PARTICIPANTS JOINTLY ADOPTED BY A COMMITTEE OF THE AMERICAN BAR ASSOCIATION AND A COMMITTEE OF PUBLISHERS.

LIBRARY OF CONGRESS CATALOGING-IN-PUBLICATION DATA

The Commodity Futures Trading Commission / Jason H. Burns, editor.
   p. cm.
"This is an excerpted, edited and indexed version of a GAO report."--Preface.
ISBN 978-1-60456-296-5
  1. United States. Commodity Futures Trading Commission. 2. Commodity futures--Law and legislation--United States. 3. Petroleum products--Prices--United States. I. Burns, Jason H. II. United States. Government Accountability Office. Commodity Futures Trading Commission.
  KF1085.A25   2007
  346.73'0922--dc22   2007048706

Published by Nova Science Publishers, Inc. ✦ New York

CONTENTS

Preface		vii
Abbreviations		ix
Chapter 1	Trends in Energy Derivatives Markets Raise Questions About CFTC's Oversight	1
Chapter 2	Results in Brief	5
Chapter 3	Background	9
Chapter 4	Several Factors Have Caused Changes in the Energy Markets, Potentially Affecting Prices	19
Chapter 5	CFTC Oversees Exchanges and Has Limited Authority over Other Derivatives Markets	31
Chapter 6	CFTC Engages in Surveillance Activities and Enforcement Activities, but the Effectiveness of These Activities is Largely Uncertain	43
Chapter 7	Conclusion	55
Appendix I:	Scope and Methodology	59
Appendix II:	Types of Contracts and Transactions for Energy Commodities in the Physical and Financial Markets	63
Appendix III:	New York Mercantile Exchange Surveillance and Enforcement Activities	65

Appendix IV:	Commodity Futures Trading Commission's Energy-Related Enforcement Actions, August 2001 - September 2006	**71**
Additional Reading		**77**
References		**79**
Index		**89**

PREFACE

This new book focuses on four energy commodities—crude oil, unleaded gasoline, natural gas, and heating oil— and Commodity Futures Trading Commission's oversight of these commodities. Specifically, this report examines (1) trends and patterns of trading activity in the physical and energy derivatives markets and the effects of those trends on prices; (2) the scope of CFTC's authority for protecting market users from fraudulent, manipulative, and abusive practices in the trading of energy futures contracts; and (3) the effectiveness of CFTC's monitoring and detection of market abuses in energy futures markets and in connection with energy-related enforcement actions. This is an excerpted, edited and indexed version of a GAO report.

ABBREVIATIONS

BCC	business conduct committee
CEA	Commodity Exchange Act
CFMA	Commodity Futures Modernization Act of 2000
CFTC	Commodity Futures Trading Commission
COMEX	New York Commodity Exchange
COT	Commitment of Traders
CPO	commodity pool operator
CTA	commodity trading advisor
DOJ	Department of Justice
EIA	nergy Information Administration
FERC	Federal Energy Regulatory Commission
FTC	Federal Trade Commission
FTE	full-time-equivalent
HU	YMEX gasoline contract
ICE	ntercontinentalExchange
ISS	ntegrated surveillance system
LTRS	large trader reporting system
NYMEX	New York Mercantile Exchange, Inc.
OMB	Office of Management and Budget
OPEC	Organization of the Petroleum Exporting Countries
OTC	over-the-counter
PART	Program Assessment Rating Tool
RB	YMEX reformulated gasoline blendstock contract
SEC	Securities and Exchange Commission
SRO	self-regulatory organization

Chapter 1

TRENDS IN ENERGY DERIVATIVES MARKETS RAISE QUESTIONS ABOUT CFTC'S OVERSIGHT

WHAT GAO FOUND

Rising energy prices have been attributed to a variety of factors, among them recent trends (2002-2006) in the physical and futures markets. These trends include (1) factors in the physical markets, such as tight supply, rising demand, and a lack of spare production capacity; (2) higher than average, but declining, volatility (a measure of the degree to which prices fluctuate over time) in energy futures prices for crude oil, heating oil, and unleaded gasoline; and (3) growth in several key areas, including the number of noncommercial participants in the futures markets (including hedge funds), the volume of energy futures contracts traded, and the volume of energy derivatives traded outside of traditional futures exchanges. Because these changes took place concurrently, the effect of any individual trend or factor is unclear.

On the basis of its authority under the Commodity Exchange Act (CEA), CFTC focuses its oversight primarily on the operations of traditional futures exchanges, such as the New York Mercantile Exchange, Inc. (NYMEX), where energy futures are traded. Energy derivatives are also traded on other markets, namely, exempt commercial and over-the-counter (OTC) markets, that are exempt from CFTC oversight. Both types of markets have seen their volumes climb in recent years. Exempt commercial markets are electronic trading facilities where certain commodities, such as energy, are traded between large, sophisticated participants. OTC markets allow eligible parties to enter into contracts directly, without using an exchange. While the exempt commercial and OTC markets are subject to the CEA's antimanipulation and antifraud provisions and CFTC enforcement of those provisions, some market observers question whether CFTC needs broader authority to oversee these markets. CFTC is currently examining the effects of trading in the regulated and exempt energy markets on price discovery and the scope of its authority over these markets—an issue that will warrant further examination as part of the CFTC reauthorization process. Moreover, because of changes and innovations in the market, the methods used to categorize these data can distort the information reported to the public, which may not be completely accurate or relevant.

CFTC conducts daily surveillance of trading on NYMEX that is designed to detect and deter fraudulent or abusive trading practices involving energy futures contracts. To detect abusive practices, such as potential manipulation, CFTC uses various information sources and relies heavily on trading activity data for large market participants. Using this information, CFTC staff may pursue alleged abuse or manipulation. However, because the agency does not maintain complete records of all such allegations, this lack of information makes it difficult to determine the usefulness and extent of these activities. In addition, CFTC's performance measures for enforcement do not fully reflect the program's goals and purposes, which could be addressed by developing additional outcome-based performance measures that more fully reflect progress in meeting the program's overall goals.

The price of energy commodities—crude oil, unleaded gasoline, heating oil, and natural gas—increased significantly from 2002 to 2006, negatively affecting consumers and the U.S. economy. While increased energy prices generally are attributed to normal market forces of supply and demand, some observers have questioned whether trading activity in energy futures contracts and other types of energy derivatives placed upward pressure on prices during this period.[1] A futures contract is an agreement to purchase or sell a commodity for delivery in the future.[2] Like other types of derivatives, its price is based on the value of an underlying commodity, such as natural gas or oil. While futures prices are determined on the basis of prices in the market where physical goods and commodities are sold (physical market), buyers and sellers of natural gas, crude oil, gasoline, and other energy products are influenced by the futures prices of these commodities when determining their prices. Trading in futures contracts has grown significantly since 2001, in part because of trading by new market participants, such as hedge funds, and increased investment in commodity index funds.[3]

The surge in energy prices and the growth in the volume of futures contracts and other derivatives have renewed questions about the adequacy of the Commodity Futures Trading Commission's (CFTC) authority and ability to oversee derivatives that are traded off exchange, or over the counter (OTC). CFTC's primary mission includes preserving the integrity of the futures markets and protecting market users and the public from fraud, manipulation, and abusive trading practices.[4] In 2000, CFTC's authority regarding futures contracts and other types of derivatives was clarified by the Commodity Futures Modernization Act of 2000 (CFMA). Among other things, the CFMA specifically authorizes off-exchange derivatives trading by establishing a framework that tailors the level of regulation of a market to the products being traded and the market's participants. Under the act, some exchanges (e.g., the New York Mercantile Exchange, Inc. (NYMEX)), that allow all types of traders, including retail customers, to access their facilities are regulated, while other venues that are off exchange can be accessed only by large, sophisticated traders and are either largely unregulated or exempt from regulation. Like futures markets, these unregulated, off-exchange markets also have grown significantly, raising questions about the amount of regulatory scrutiny that CFTC should provide.

This report, conducted under the Comptroller General of the United States' authority, addresses concerns raised by Congress, consumer groups, states' attorneys general, and others about rising prices in energy markets and the relationship, if any, of futures trading to rising energy prices. We addressed this report to you because of your expressed interest or your committee's jurisdiction. This report focuses on four energy commodities—crude oil, unleaded gasoline, natural gas, and heating oil—and CFTC's oversight of these commodities. Specifically, this report examines (1) trends and patterns of trading activity in the physical and energy derivatives markets and the effects of those trends on prices; (2) the scope of CFTC's authority for protecting market users from fraudulent, manipulative, and abusive practices in the trading of energy futures contracts; and (3) the effectiveness of CFTC's monitoring and detection of market abuses in energy futures markets and in connection with energy-related enforcement actions.

To address these objectives, we obtained and analyzed end-of-the-day trading data for energy futures contracts from NYMEX and data from CFTC's large trader reporting system (LTRS) database, which we tested and found reliable for our purposes.[5] We obtained and analyzed other CFTC records and reports relevant to the commission's surveillance and other activities. We also reviewed applicable laws, regulations, and policy statements. We obtained information from a broad range of participants in the energy futures markets and officials knowledgeable about the futures markets. These individuals included officials from large oil companies, refiners, trade associations representing end users of natural gas, investment banks, and hedge funds as well as energy consultants and academic experts. We interviewed officials in CFTC's Division of Market Oversight, Division of Enforcement, Office of the Chief Economist, Office of the General Counsel, and Office of the Inspector General. Moreover, because CFTC oversight is also provided through officials located in the commission's field offices, we obtained information from officials at the CFTC New York Regional Office, which conducts surveillance of futures trading on NYMEX. In addition, we gathered and analyzed information on oversight of the energy markets provided by other federal agencies, including the U.S. Department of Energy's Energy Information Administration (EIA), the Federal Energy Regulatory Commission (FERC), the Federal Trade Commission (FTC), the Department of Justice (DOJ), and the Securities and Exchange Commission (SEC). We conducted our work in Chicago, Houston, New York City, and Washington, D.C., between July 2005 and September 2007 in accordance with generally accepted government auditing standards. Appendix I contains a more detailed description of our scope and methodology.

Chapter 2

RESULTS IN BRIEF

Significant changes occurred in both physical and energy derivatives markets between 2002 and 2006 that were accompanied by rising energy prices; however, it is difficult to precisely determine the extent or effect of any single factor on energy prices. Specifically:

- There was a tight supply and rising demand in the physical markets for crude oil, heating oil, unleaded gasoline, and natural gas, stemming from various factors—such as increased political instability in some of the major oil-producing countries, decreased spare oil production capacity, refining capacity that did not expand at the same pace as demand for gasoline, and rapidly rising global demand for energy products.
- Volatility (a measure of the degree to which prices fluctuate over time) in energy futures prices generally remained above historic averages in 2002 and 2003, but declined through 2006 for crude oil, heating oil, and unleaded gasoline.
- The number of noncommercial participants in the futures markets, the volume of energy futures contracts traded, and the volume of energy derivatives traded outside traditional futures exchange also have grown steadily.

Reasonable arguments have been made that events in physical and futures markets contributed in some degree to the increases in inflation-adjusted energy prices in both markets during this period for crude oil, unleaded gasoline, and heating oil. However, opinions vary on how much the recent changes in the financial markets influenced energy prices. For example, some market participants and observers have argued that speculation alone could not have influenced prices artificially over such a long period, while others have concluded that increased trading activity put upward pressure on the prices of spot as well as futures contracts.

Under the Commodity Exchange Act (CEA), CFTC's authority for protecting market users from fraudulent, manipulative, and abusive practices in energy derivatives trading is primarily focused on the operations of traditional futures exchanges, such as NYMEX, where energy futures are traded. To help provide transparency to the public, CFTC publishes aggregate trading information for large commercial (such as oil companies and refineries) and noncommercial (such as hedge funds) traders for various commodities through its Commitment of Traders (COT) reports. These reports include the number of traders, changes since the last report, and open positions—an obligation to take or make delivery of a commodity in the future without a matching obligation in the opposite direction. However, because of changes and innovation in the market, methods used to categorize these data can distort the accuracy and relevance of the information reported to the public. The market for energy derivatives also has changed in other ways. Specifically, trading has grown on other markets, namely, exempt commercial markets—electronic trading facilities that trade exempt commodities, more than half of which trade in energy products—and OTC markets.[6] Currently, CFTC receives limited information on derivatives trading on exempt commercial markets—for example, records of allegations or complaints of suspected fraud or manipulation, and price, quantity, and other data on contracts that average five or more trades a day. The commission may receive limited information from OTC participants, such as trading records, to help CFTC enforce the CEA's antifraud or antimanipulation provisions. The scope of CFTC's oversight authority with respect to these markets has raised concerns among some Members of Congress and others that activities on these markets are largely unregulated, and that additional CFTC oversight is needed. While many regulators have resisted calls for more regulation in the past, recent events in the physical and energy derivatives markets have resulted in renewed focus on the sufficiency of CFTC's authority. As a result, CFTC held a hearing in September 2007 to begin examining trading on regulated exchanges and exempt commercial markets. The hearing included assessments of the relationship between these markets and assessments of whether markets other than NYMEX serve a price discovery function, which is the process of determining a commodity's price on the basis of supply and demand. These and future deliberations may provide insights into whether changes are needed in the scope of CFTC's authority. Depending on what

CFTC finds in its assessments of the markets, Congress might want to consider what actions, if any, are warranted.

To detect fraudulent or abusive trading practices involving exchange-traded energy futures, CFTC daily monitors the trading on exchanges such as NYMEX. CFTC examines daily electronic trading data on futures contracts and other information sources, such as commercial sources on energy commodities and tips from individuals on possible violations. CFTC's surveillance program primarily relies on daily reports from large traders to detect problems, such as the potential for manipulation. When CFTC staff detect potential problems or violations, they may gather additional information from NYMEX officials, traders, or other sources to determine if further action is warranted. CFTC staff said that they routinely investigated traders with large open positions. However, the staff added that they did not routinely maintain information about such inquiries; instead they documented their actions only when further action was warranted. This lack of information makes it difficult to determine the usefulness and extent of these activities. Without sufficient data on these and other inquiries, CFTC's records will understate the extent to which the commission surveils trading activity. In addition, CFTC management also might miss opportunities both to identify trends in activities or markets and to better target its limited resources. According to information provided by CFTC, the commission coordinates its enforcement actions with NYMEX as well as FERC, DOJ, and others. It also has taken enforcement actions in cases of attempted manipulation and other abusive practices in energy derivatives trading that resulted in fines of $305 million from 2001 through 2005. While these cases have been successfully pursued, it is difficult to determine whether they have helped deter market manipulation or the other abusive practices these pursuits addressed because the effectiveness of enforcement activities is not easily measured. The Office of Management and Budget (OMB) has concluded that the enforcement program lacks performance measures that illustrate whether it is meeting its overall objective.

This report includes a matter for congressional consideration and three recommendations. In light of recent developments in derivatives markets and as part of CFTC's reauthorization process, Congress should consider further exploring whether the current regulatory structure for energy derivatives, in particular for those traded in exempt commercial markets, provides adequately for fair trading and accurate pricing of energy commodities. Our three recommendations to the Acting CFTC Chairman are aimed at improving the usefulness of information that CFTC provides to the public as a result of its surveillance activities and the efficiency of its enforcement program. First, we recommend that CFTC reexamine the classifications in the COT reports to determine if the commercial and noncommercial categories should be refined to improve the transparency, accuracy, and relevance of public information on trading activity in the energy futures markets. Second, we recommend that CFTC explore ways to routinely maintain written records of inquiries into possible improper trading activity and the results of these inquiries to more fully determine the usefulness and extent of its surveillance, antifraud, and antimanipulation authorities. Third, we recommend that CFTC examine ways to more fully demonstrate the effectiveness of its enforcement activities by developing additional outcome-related performance measures that more fully reflect progress on meeting the program's overall goals.

We provided a draft of this report to CFTC, and the commission provided written comments that are reprinted in appendix V. In its comments, CFTC generally agreed with our findings. CFTC said that the commission will reexamine classifications in the COT reports. CFTC also said that the commission will explore additional recordkeeping procedures for staff, but that it must balance the time required for such additional tasks against the need to undertake market surveillance by an already-stretched surveillance staff. CFTC added that it has included the development of measures to evaluate the effectiveness of its enforcement program in its most recent strategic plan. CFTC also provided technical comments, which we have incorporated in this report as appropriate.

Chapter 3

BACKGROUND

Energy commodities are bought and sold in several different physical and financial markets. Physical markets include the spot, or cash, markets where products such as crude oil or gasoline are bought and sold for immediate or near-term delivery. The United States has several spot markets. Examples are the pipeline hub near Cushing, Oklahoma for West Texas Intermediate crude oil and the Henry Hub near Erath, Louisiana, for natural gas. The prices set in the specific spot markets provide a reference point that buyers and sellers use to set the price for other types of the commodity traded in other locations.

The prices established for energy commodities in the physical markets generally are determined by supply and demand. For example, when the demand for the product rises relative to supply because economies are growing, prices are likely to rise. Conversely, when demand falls relative to supply, prices are likely to fall. For energy products, demand and supply, and therefore price, can fluctuate on a seasonal basis. For example, consumer demand for gasoline in the United States is generally higher from May through early September—the summer driving season—and tends to flatten after Labor Day. Similarly, demand for natural gas and heating oil is highest during the heating season between October and March.

The relative inelasticity of energy commodities means that small shifts in demand and supply can result in relatively large price fluctuations. In general, when the price of an energy commodity rises, the demand for that product is likely to fall in the long term, and vice versa. However, demand for energy commodities is price inelastic in the short term—that is, the quantity demanded changes little in response to a change in price. On the supply side, rising energy commodities prices motivate producers to increase the amount of commodities they supply to increase profits.

However, because producers hold relatively low inventories of energy commodities in reserve, and finding and producing additional energy commodities takes a long time and is expensive, supply also is relatively inelastic. For example, supplies of natural gas from new production wells cannot be increased quickly to meet higher demand because of the time required to get the newly produced gas into the marketplace.

Energy commodities also are traded in the financial markets, especially in the form of derivatives. Derivatives include futures, options, and swaps, whose values are based on the performance of the underlying asset. Options give the purchaser the right, but not the obligation, to buy or sell a specific quantity of a commodity or financial asset at a designated price. Swaps traditionally are privately negotiated contracts that involve an ongoing exchange of one or more assets, liabilities, or payments for a specified period. Futures and options contracts are traded on exchanges designated by CFTC as contract markets (futures exchanges), where a wide range of energy, agricultural, financial, and other commodities are bought and sold for future delivery. Commodity futures and options can be traded on both OTC and exempt commercial markets if the transactions involve qualifying commodities and the participants satisfy statutory requirements.

Energy futures include standardized contracts for future delivery of a specific crude oil, heating oil, natural gas, or gasoline product at a particular spot market location. The exchange standardizes the contracts, and participants cannot modify them to their particular needs. For example, a standard gasoline futures contract traded on NYMEX is for 1,000 barrels (42,000 gallons), quoted in dollars and cents per gallon, and for delivery of up to 36 months into the future at New York Harbor.[7] The owner of an energy futures contract is obligated to buy or sell the commodity at a specified price and future date. However, the owner may eliminate the contractual obligation before the contract expires by selling or purchasing other contracts with terms that offset the original contract. In practice, relatively few futures contracts on NYMEX result in physical delivery of the underlying commodity, but instead are liquidated with offsets. Options on futures contracts also are traded on exchanges such as NYMEX and foreign boards of trade that U.S. traders access directly.

In addition to exchange-traded futures and options, the financial markets for energy commodities include derivatives traded among multiple traders on exempt commercial markets and derivatives created bilaterally in OTC transactions. As with futures, exempt commercial markets and other OTC derivatives allow producers and users of energy commodities to manage the risk of future changes in the price of a particular commodity. These contracts include options and swaps at an agreed-upon price. Appendix II shows some of the different types of contracts and transactions for energy commodities in the physical and financial markets.

FUNCTIONS OF FUTURES MARKETS

Market participants use futures markets to offset the risk caused by changes in prices, discover commodity prices, and speculate on price changes. Some buyers and sellers of energy commodities in the physical markets trade in futures contracts to offset, or "hedge," the risks of price changes in the physical markets. The futures markets help buyers and sellers determine, or "discover," the price of commodities in the physical markets, thus linking the two markets. Other participants—generally, speculators—that do not have a commercial interest in the underlying commodities but are looking to make a profit take varying positions on the future value of commodities. In doing so, speculators provide liquidity and assume risks that other participants, such as hedgers, seek to avoid. Arbitrageurs are a third group of participants that aim to benefit by identifying discrepancies in price relationships, rather than by betting on future price movements. Arbitrage is a strategy that involves simultaneously entering into several transactions in multiple markets to benefit from price discrepancies across markets. For example, traders can trade simultaneously in exchanges and OTC.

Price risk is an important concern for buyers and sellers of energy commodities because wide fluctuations in cash market prices introduce uncertainty for producers, distributors, and consumers of commodities and make investment planning, budgeting, and forecasting more difficult. A statistical measurement of the degree to which prices fluctuate over time is known as "volatility" and can be applied to prices in both the physical and financial markets. There are two basic types of volatility measurements. Historical volatility measures are calculated on the basis of price changes, using data from market transactions. Implied volatility reflects market participants' expectations of future volatility as derived from the prices of traded options (see app. III). This report presents data on the relative historical volatility of energy futures contracts, which we calculated from relative changes in daily prices.

Futures and off-exchange derivatives markets provide participants with a means to hedge or shift unwanted price risk to others more willing to assume the risk or those having different risk situations. For example, if a petroleum refiner wanted to shed its risk of losing money as a result of falling gasoline prices, it could lock in a price by selling futures contracts to deliver the gasoline in 6 months at a guaranteed price. Likewise, a transportation company that knows it must refill its gasoline tanks in 6 months might want to offset the price risk associated with purchasing fuel by buying futures contracts to take delivery of gasoline then at a set price. Without futures contracts that help them manage risk, producers, refiners, and others likely would face uncertainty related to investment planning, budgeting, and forecasting—and potentially higher costs.

Futures markets also provide a means of price discovery for commodities such as energy products. For price discovery, markets need current information about supply and demand, a large number of participants, and transparency. Market participants monitor and analyze the factors that currently affect, and that they expect to affect, the future supply and demand for energy commodities. With that information, they buy or sell energy commodity contracts on the basis of the price for which they believe the commodity will sell at the delivery date. The futures markets, in effect, distill the diverse views of market participants into a single price. In turn, buyers and sellers of physical commodities consider those predictions about future prices with other factors when setting prices on the spot and retail markets.

A wide variety of participants hedge and speculate in energy derivatives markets. For the exchange-traded futures markets, CFTC categorizes traders in general terms as either commercial or noncommercial participants. CFTC identifies several subcategories of participants within the commercial category: producers, manufacturers, dealers/merchants, and swaps/derivatives dealers. Dealers and merchants include, among others, wholesalers, exporters and importers, shippers, and crude oil marketers. Typical noncommercial traders are entities such as those that manage money ("managed money traders").[8] These noncommercial traders include, among others, commodity pool operators (CPO) and commodity trading advisors (CTA), many of which advise or operate hedge funds.[9] Other noncommercial traders include floor brokers and unregistered traders.

RELATIONSHIP BETWEEN FUTURES AND SPOT PRICES

The prices for energy commodities in the futures and in the spot or physical markets are closely linked because they are influenced by the same market fundamentals in the long run. Prices in the physical spot and futures markets for the four energy commodities we reviewed are highly correlated and rose dramatically from 2002 to 2006. As shown in figure 1, from January 2002 to July 2006, monthly average spot prices for crude oil, gasoline, and heating oil increased by at least 220 percent.[10] Natural gas spot prices increased by more than 140 percent. At the same time that spot prices increased, the futures prices for these commodities showed a similar pattern of a sharp and sustained increase from January 2002 into 2006. For example, the price of crude oil futures increased from an average of $22 per barrel in January 2002 to an average of $74 per barrel in July 2006. Natural gas futures prices spiked rapidly in the fall of 2005 after several strong hurricanes raised concerns about supply disruptions for the winter of 2005-2006, then prices fell sharply due in part to a mild winter. Prices in the spot and futures markets show similar patterns because traders in those markets tend to rely on the same types of information when entering into transactions.

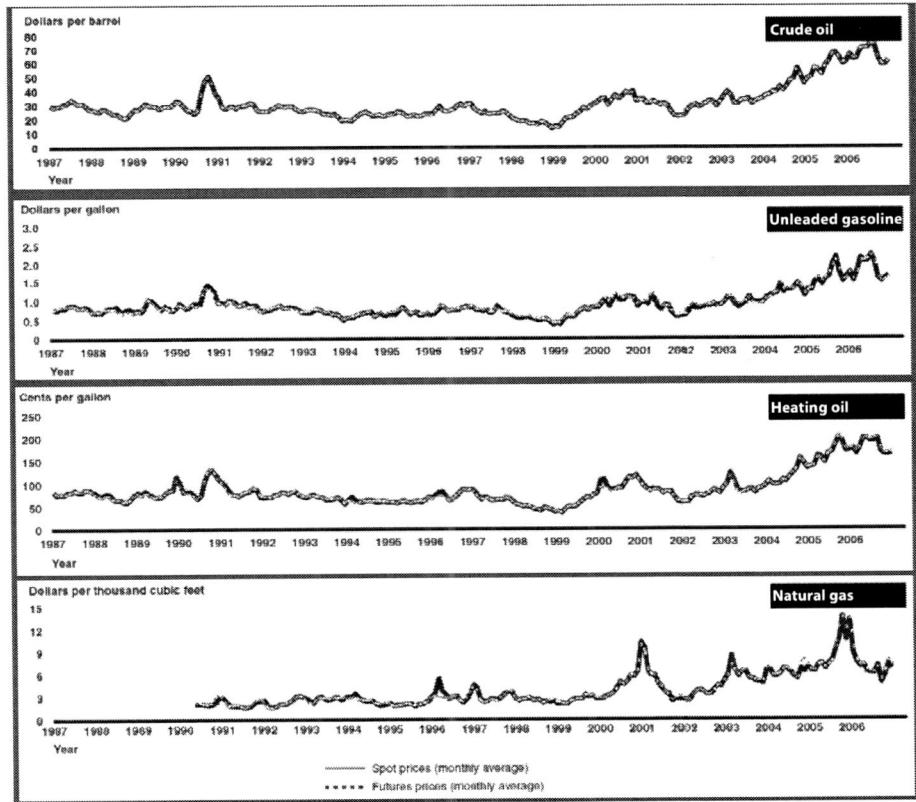

Source: GAO analysis of Global Insight and NYMEX data.
Note: The front month futures contract is the actively traded contract with the closest delivery date. NYMEX did not begin trading natural gas futures until 1990.

Figure 1. Monthly Average Spot Prices and Front Month Futures Settlement Prices, in Constant 2006 Dollars, 1987–2006.

The differences between futures and spot market prices for energy commodities narrow and the prices converge when futures contracts near expiration and physical delivery is required. As the expiration date nears, the physical delivery provision of the contract and the ability of traders to arbitrage combine to bring the futures and physical market prices together. Arbitrage plays a crucial role in moderating or removing price differences between spot and futures markets and contributes to the convergence of futures and spot prices at expiration. For example, if the price for a crude oil futures contract that would expire in 2 weeks were $62 per barrel and the spot market price were $60 per barrel, a trader could choose to buy oil now at the spot price and enter into a futures contract to deliver oil in 2 weeks at the futures price, thereby making a $2 profit.[11] This and similar transactions by other traders would put upward pressure on the spot price and downward pressure on the futures price and move them toward convergence. Figure 2 provides an example of how the price of the April 2006 crude oil futures contract and the spot price for that commodity converged as the contract approached expiration.

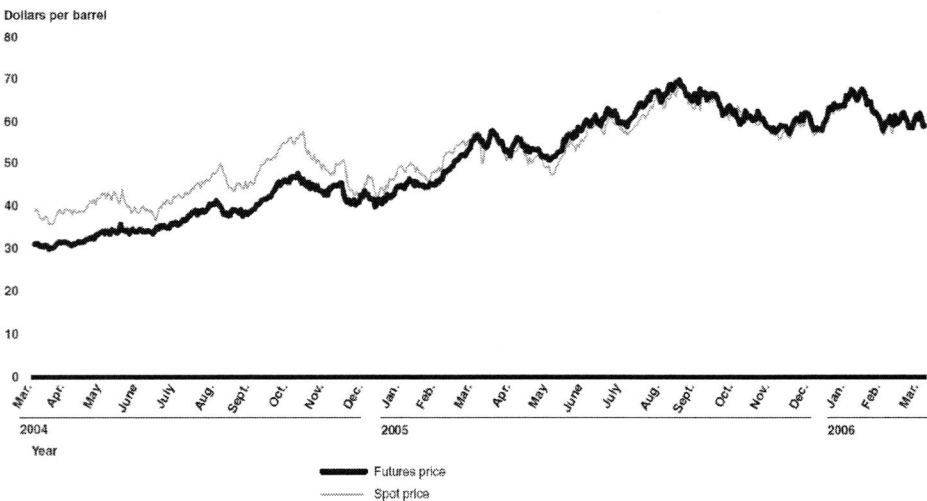

Source: GAO analysis of EIA and NYMEX data.

Figure 2. Convergence of the April 2006 Crude Oil Futures Contract Price and the Crude Oil Spot Price, March 22, 2004–March 21, 2006.

CHANGES IN CFTC OVERSIGHT AUTHORITY AND RESOURCE LEVELS

Between the creation of CFTC in 1974 and the year 2000, the CEA generally restricted commodity derivatives trading to futures and options entered into on exchanges and made all transactions in futures contracts subject to CFTC's exclusive jurisdiction. However, in the late 1980s and early 1990s, commercial entities began entering into nonstandardized, off-exchange derivative contracts that had pricing characteristics similar to futures (i.e., pricing of the transactions derived from the prices of various commodities), and the instruments were used for risk shifting. According to CFTC officials, under exemptive authority provided in 1992 reauthorization legislation, CFTC announced that it would not take enforcement action against qualified commercial entities engaged in certain types of energy derivatives transactions, but the legality of instruments not covered by the exemption (i.e., their status as futures contracts subject to the CEA) remained unresolved.[12]

In 2000, the CFMA amended the CEA to provide for both regulated markets and markets largely exempt from regulation and to permit off-exchange trading of energy derivatives by qualified parties.[13] The regulated markets include futures exchanges that have self-regulatory surveillance and monitoring responsibilities as self-regulatory organizations (SRO) and by CFTC.[14] CFTC's primary mission includes preserving the integrity of these futures markets and protecting market users and the public from fraud, manipulation, and abusive practices related to the sale of commodity futures and options. This mission is achieved through a regulatory scheme that is based on federal oversight of industry self-regulation. The CEA also permits derivatives trading in markets that are largely exempt from CFTC's regulatory authority, including both OTC and exempt commercial markets, subject to statutory requirements governing the types of commodity and trader and the facility used for conducting the trades. The President's Working Group's 1999 report on OTC derivatives focused on changes to the CEA that in their view would "promote innovation, competition, efficiency, and transparency in OTC derivatives markets, to reduce systemic risk, and to allow the United States to maintain leadership in these rapidly developing markets."[15] Derivatives on energy commodities, which are within the act's definition of "exempt commodity," may be traded in exempt commercial markets by eligible commercial entities, a category of traders broadly defined in the CEA to include firms with a commercial interest in the underlying commodity as well as other sophisticated investors, such as hedge funds. Violations of the CEA and CFTC regulations may be remedied by imposition of civil monetary penalties, trading bans, restitution, and other appropriate relief.

In addition to CFTC oversight, futures exchanges accept self-regulatory obligations as a condition of designation. For example, NYMEX, as an SRO, is responsible for establishing and enforcing rules governing member conduct and trading; providing for the prevention of market manipulation, including monitoring trading activity; ensuring that futures industry professionals meet qualifications; and examining exchange members for financial soundness and other regulatory purposes. CFTC oversees SROs to ensure that each has an effective self-regulatory program.[16]

Within CFTC, three of the commission's six major operating units actively oversee futures exchanges and their derivatives clearing organizations.[17]

- The Division of Market Oversight approves and oversees the futures exchanges, conducts its own market surveillance, conducts trade practice reviews and investigations, and reviews exchange rules.
- The Division of Clearing and Intermediary Oversight oversees, among other things, derivatives clearing organizations and the registration of intermediaries, which are persons such as futures commission merchants, CPOs, or CTAs that act on the behalf of others in futures trading.[18]
- The Division of Enforcement investigates and prosecutes alleged violations of the CEA and CFTC regulations.

At the beginning of fiscal year 2006, 167 (34 percent) of CFTC's 490 full-time-equivalent (FTE) positions were allocated to the first two CFTC divisions; at the beginning of fiscal year 2007, that allocation declined to 162 (35 percent) of CFTC's 458 FTE positions. These staff monitor the markets and market participants from CFTC's headquarters in Washington, D.C., as well as from field offices in New York; Chicago; Kansas City; and, until recently, Minneapolis.[19] About one-third of CFTC's staff are located in the field offices. At the beginning of fiscal year 2006, 132 (27 percent) of CFTC's 490 FTE positions were allocated to the Division of Enforcement; at the beginning of fiscal year 2007, that number declined to 120 of CFTC's 458 FTE positions. The 2007 data are estimated. While CFTC staffing levels have declined, according to CFTC, futures and options trading volume for all commodities has roughly doubled from fiscal years 2002 to 2006 and is expected to continue to rise, as indicated in figure 3.

Background

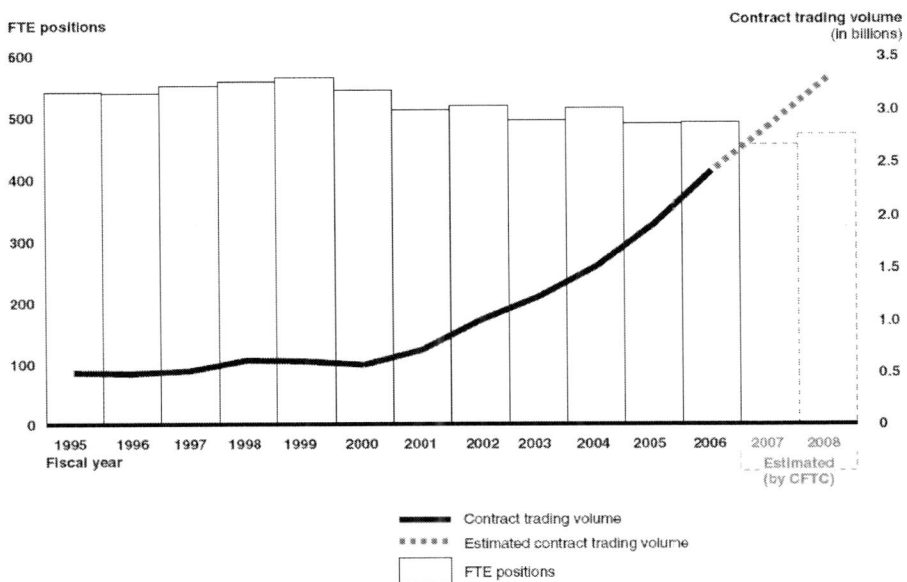

Source: GAO analysis of CFTC, the National Finance Center, and the Futures Industry Association data.

Figure 3. Futures and Options Trading Volume for All Commodities and CFTC Staffing Levels (Actual and Estimated), Fiscal Years 1995–2008.

Chapter 4

SEVERAL FACTORS HAVE CAUSED CHANGES IN THE ENERGY MARKETS, POTENTIALLY AFFECTING PRICES

Both physical and futures markets experienced a substantial amount of change from 2002 through 2006. Reasonable arguments have been made that events in both markets have contributed to rising energy prices, at least in the short term, but opinions vary regarding the extent that recent changes in the financial markets have influenced the prices of energy products in the physical markets over the long term. Because of these concurrent changes, identifying the causes of the increases in energy prices in both the physical and futures markets for crude oil, unleaded gasoline, heating oil, and natural gas is difficult. First, during this period, the physical markets experienced tight supply and rising demand from increasing global demand, ongoing political instability in oil-producing regions, and other supply disruptions. Second, annual volatility of energy prices remained above historic averages during the beginning of the period (although during 2006, volatility generally declined to levels at or near the historical average). Third, the volume of trading in energy futures increased as growing numbers of managed money traders viewed energy futures as attractive investment alternatives.

TIGHT SUPPLY AND RISING DEMAND FOR PHYSICAL ENERGY COMMODITIES CONTRIBUTED TO THE INCREASE IN FUTURES AND SPOT PRICES

The energy physical markets have undergone substantial change and turmoil from 2002 through 2006, which affected prices in the spot and futures markets. First, like many market observers and participants, we found a number of fundamental supply and demand conditions that could influence prices. Moreover, these parties have observed that the lack of spare capacity in certain areas, such as production, transportation, and storage, can affect prices. Second, over the short term, weather events also were a significant cause of rising energy prices because of their effects on energy supply, according to several of the market observers we interviewed. Third, many market observers also identified geopolitical uncertainty arising from the instability and insecurity of the world's major oil-producing regions as a major factor affecting energy prices.[20] Concerns about political events may manifest in the form of higher futures prices if traders predict that an event—such as a strike within the industry or pipeline sabotage by terrorists—will have an effect on future supply. Finally, on the demand side, a significant factor noted by observers was the increase in global consumption of petroleum products, primarily among industrializing Asian nations such as China and India.

Analysis of world oil prices by EIA and us indicates that increases in crude oil prices occur if political instability, terrorist acts, or natural disasters create uncertainties about, or actual disruptions in, supply from countries that produce or refine oil. For example, according to EIA, in the early 2000s, cutbacks in the Organization of the Petroleum Exporting Countries (OPEC) production and rising demand caused oil prices to increase to more than $30 per barrel, only to fall precipitously when the global economy weakened following the September 11, 2001, crisis.[21] Moreover, as we reported in 2005, rapid growth in oil demand in Asia contributed to a rise in crude oil prices to more than $50 per barrel during 2004.[22]

According to EIA, world oil demand that was about 59 million barrels per day in 1983 grew to more than 85 million barrels per day in 2006. The United States consumes nearly one-quarter of this amount—or more than 20 million barrels per day in 2006—and its demand has grown about 1.5 percent per year since 1983. The rapid economic growth in Asia also has stimulated a strong demand for energy commodities. For example, China has overtaken Japan as the second-largest consumer of crude oil, after the United States. According to EIA data, from 1983 to 2004, Chinese demand grew from about 1.7 million barrels consumed per day to about 6.4 million barrels consumed per day. This increase in the global demand for crude oil is shown in figure 4.

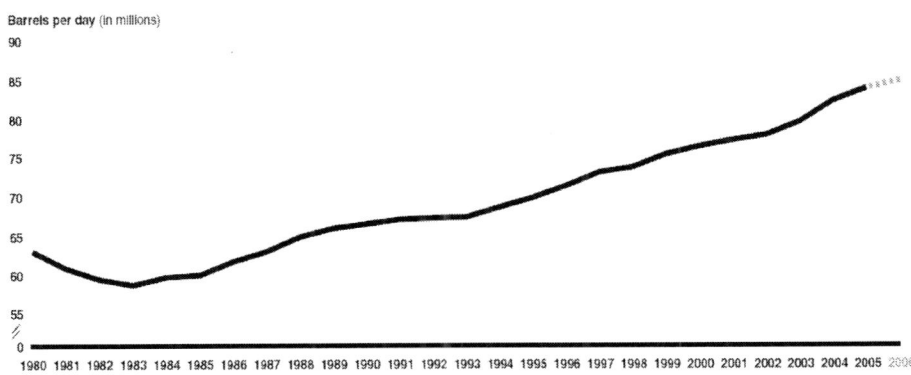

Source: GAO analysis of EIA data.
Note: The world oil demand data for 2006 represent a preliminary estimate.

Figure 4. Increase in World Demand for Crude Oil (Actual and Estimated), 1980–2006.

The growth in demand does not, by itself, lead to higher prices for crude oil or any other energy commodity. For example, if the growth in demand were exceeded by a growth in supply, prices would fall, with other things remaining constant. However, according to EIA, the growth in demand outpaced the growth in supply, even with spare production capacity included in supply. Spare production capacity is surplus oil that can be produced and brought to the market relatively quickly to rebalance the market if there were a supply disruption anywhere in the world oil market. EIA estimates that global spare production capacity in 2006 was about 1.3 million barrels per day (see figure 5). Most of that capacity was concentrated in the 12 OPEC countries that supply about 40 percent of the world's oil, primarily Saudi Arabia. This compared with spare capacity of about 10 million barrels per day in the mid-1980s, or of about 5.6 million barrels a day as recently as 2002. Analysis by EIA indicates that the growth of oil production in non-OPEC nations, which produce most of the world's oil and include countries such as Canada, China, Mexico, Norway, Russia, the United Kingdom, and the United States, has slowed relative to the growth in demand, and these nations have virtually no spare production capacity. As a commodity that is produced and traded worldwide, crude oil prices could be affected by the value of the U.S. dollar on open currency markets. For example, because crude oil is typically denominated in U.S. dollars, the payments that oil-producing countries receive for their oil also are denominated in U.S. dollars. As a result, a weak U.S. dollar decreases the value of the oil sold at a given price, and oil-producing countries may wish to increase prices for their crude oil to maintain purchasing power in the face of a weakening U.S. dollar, to the extent they can.

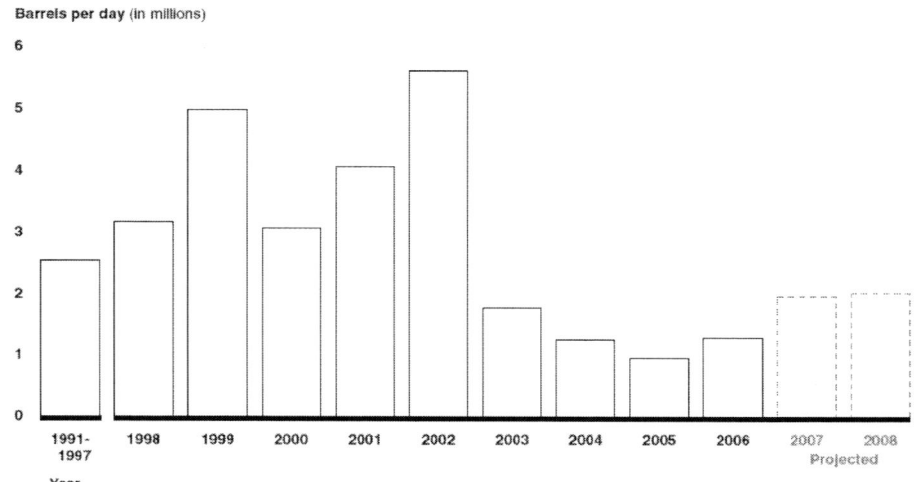

Source: GAO analysis of EIA data.
Note: The spare production capacity data for 1991–1997 represent an average estimate over that period.

Figure 5. Estimates of World Oil Spare Production Capacity, 1991–2008.

Major weather and political events also can lead to supply disruptions and higher prices. In its analysis, EIA has cited the following examples:

- Hurricanes Katrina and Rita removed about 450,000 barrels per day from the world oil market from June 2005 to June 2006.
- Instability in major OPEC oil-producing countries, such as Iran, Iraq, Nigeria, and Venezuela, has lowered production and increased the risk of future production shortfalls.
- Oil production in Russia, a major driver of non-OPEC supply growth during the early 2000s, was adversely affected by a worsened investment climate as the government raised export and extraction taxes.

The supply of crude oil affects the supply of gasoline and heating oil, and, just as production capacity affects the supply of crude oil, refining capacity affects the supply of products distilled from crude oil. As we have reported, refining capacity in the United States has not expanded at the same pace as the demand for gasoline.[23] Despite a growth in the capacity of existing gasoline refineries, the growth in demand has meant that refineries have been running at an average of more than 93 percent of production capacity since the mid-1990s, compared with about 78 percent in the 1980s. Higher utilization rates can increase operating costs and lead to prices being higher than otherwise would be expected, as occurred in the second half of the 1990s.

Another factor affecting the supply, and therefore the price, of petroleum products is the amount held in inventory. Inventory is particularly crucial to the supply and demand balance because it can provide a cushion against price spikes if, for example, a refinery outage temporarily disrupts production. We have reported that, as in other industries, the petroleum products industry has adopted "just-in-time" delivery processes to reduce costs, leading to a downward trend in the level of gasoline inventories in the United States. For example, in the early 1980s, private companies held stocks of gasoline in excess of 35 days of average U.S. consumption; while in 2004, those stocks were equivalent to less than 25 days consumption.[24] Lower costs of holding inventories may reduce gasoline prices, but lower levels of inventories also may cause prices to be more volatile because when a supply disruption occurs or there is an increase in demand, there are fewer stocks of readily available gasoline from which to draw, thereby putting upward pressure on prices. Others have noted that higher prices for future delivery of oil have induced oil companies to buy more oil and place it in storage. They concluded that this practice has created a situation where oil prices are high despite high levels of oil in inventory.

In addition to the supply and demand factors that generally apply to all energy commodities, there are specific conditions that apply to particular commodities. For example, to meet national air quality standards under the Clean Air Act, as amended, many states have mandated the use of special gasoline blends—so-called "boutique fuels." As we have recently reported, there is a general consensus that higher costs associated with supplying special gasoline blends contributed to higher gasoline prices, either because of more frequent or more severe supply disruptions or because higher costs are likely passed on, at least in part, to consumers.[25] As another example, according to EIA, the recent phaseout of a chemical used to improve gasoline performance—methyl tertiary butyl ether—increased the price of U.S. gasoline, in part because the chemical was replaced by ethanol, a more costly additive. As in the futures markets, the physical markets have undergone substantial changes that can affect prices. These specific factors affecting particular commodities, when combined with the general supply and demand conditions, contribute to increased energy prices and price volatility.

However, market participants and other observers disagree on whether high energy prices were solely due to supply and demand fundamentals or whether increased futures trading activity also was fueling higher prices.

THE EFFECTS OF RELATIVELY HIGH BUT FALLING VOLATILITY AND A GROWING VOLUME OF DERIVATIVES TRADING ON ENERGY PRICES ARE UNCLEAR

The changes occurring in the physical markets have not happened in isolation; they have been accompanied by advances in technology, relatively high but falling volatility in energy futures prices, and a growing volume of trading in the derivatives markets. The effects of these changes on energy prices are not clear.

Although energy futures prices increased from 2002 to 2006 (see figure 1), the relative volatility of those prices for three of the four commodities generally declined. As shown in figure 6, the annual historical volatilities—measured using the relative change in daily prices of energy futures—from 2000 through 2006 generally were above or near their long-term averages, although crude oil and heating oil declined below the average and gasoline declined slightly. As we have reported, futures prices typically reflect the effects of such world events on the price of crude oil.[26] Political instability and terrorist acts in countries that supply oil create uncertainties about future supplies, which is reflected in futures prices in anticipation of an oil shortage and expected higher prices in the future. Conversely, news about a new oil discovery that would increase world oil supply could result in lower futures prices. In other words, futures traders' expectations of what may happen to world oil supply and demand influence their price decisions.

The annual volatility of natural gas fluctuated more widely than that of the other three commodities and increased in 2006, even though prices largely declined from the levels reached in 2005. EIA has stated that the volatility of natural gas prices is due to factors in the physical marketplace, such as changing weather, producers' inability to move natural gas quickly to areas in response to quickly rising demand, and limited local storage. A research director for a consumer advocacy organization who studied natural gas prices concluded that increased trading by speculators had increased volatility and prices.[27] CFTC also has studied this issue and found that natural gas prices from August 2003 through August 2004 did not appear to be determined by any single category of market participant, although joint demand and supply of contracts by all participants clearly affected the change in price. In other words, managed money traders' activity (including hedge funds), by itself, did not have a significant effect on price changes.[28]

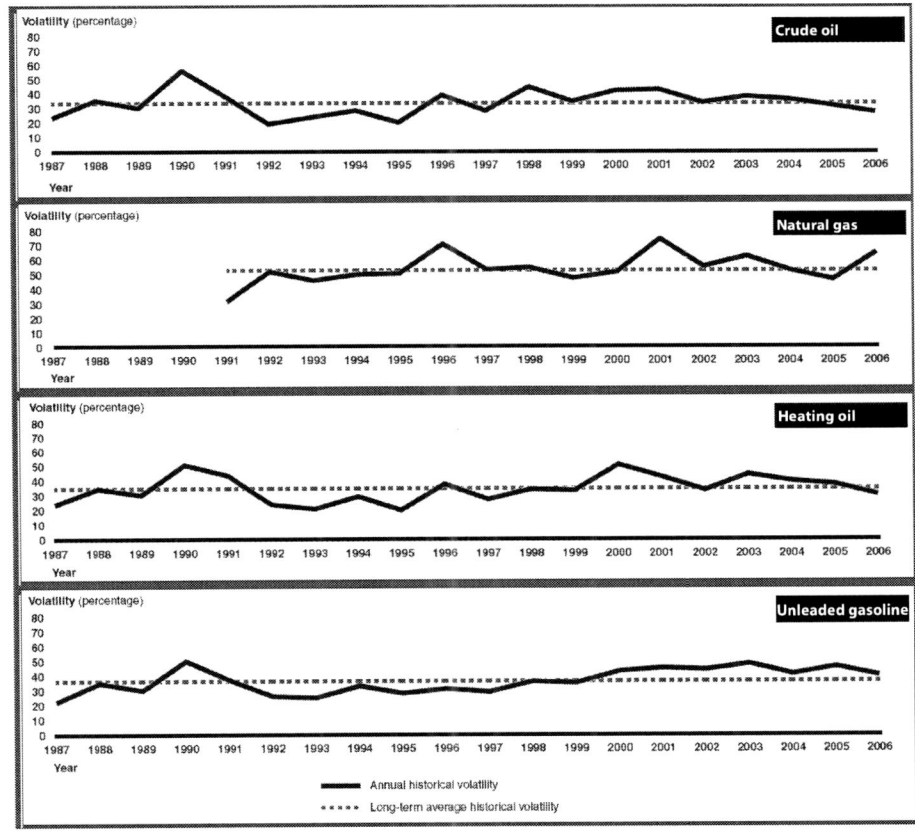

Source: GAO analysis of NYMEX data.
Note: NYMEX did not begin trading natural gas futures until 1990.

Figure 6. Comparison of Annual Volatility with the Long-term Average Volatility for Four Energy Futures (Measured in Relative Terms Using Front Month Contracts), 1987–2006.

While some often equate higher prices with higher volatility, an increase in futures contract prices does not necessarily mean that volatility will increase in a similar manner, and an increase in volatility does not necessarily mean that prices will rise. Price volatility measures the variability rather than the direction of price changes and is based on the standard deviation of those changes.[29] Therefore, if futures contract prices change at a steady rate, the prices may have lower volatility than if large swings in prices occurred.

At the same time that prices were rising and volatility was generally above or near long-term averages, futures markets also experienced an increase in the number of large noncommercial participants, such as managed money traders.[30] The trends in price and volatility made the energy derivatives markets attractive for an increasing number of traders looking to either hedge against those changes or profit from them. According to CFTC large trader data, from July 2003 to December 2006, crude oil futures and options contracts experienced the most dramatic increase as the average number of noncommercial traders grew from about 125 to about 286. As shown in figure 7, over a similar period, the average number of noncommercial traders also showed an upward but less dramatic trend for unleaded gasoline, heating oil, and natural gas.

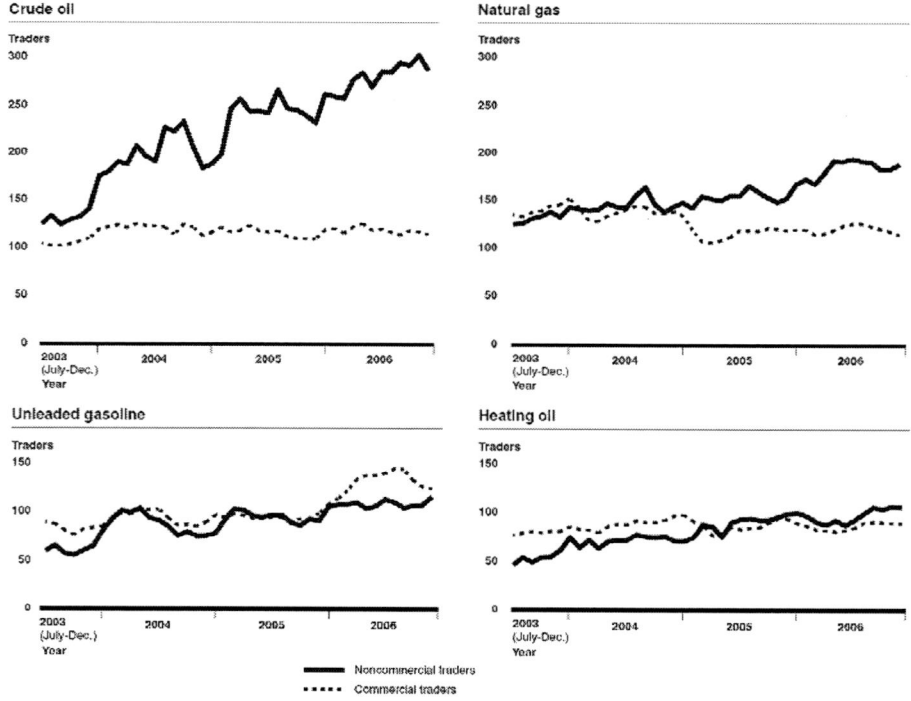

Source: GAO analysis of CFTC data.

Figure 7. Average Daily Number of Large Commercial and Noncommercial Traders per Month, July 2003–December 2006.

Some market participants and observers have concluded that large purchases of oil futures contracts by speculators in effect have created an additional demand for oil that has led to higher prices; others disagree. The Senate's Permanent Subcommittee on Investigations, Committee on Homeland Security and Governmental Affairs, issued a staff report in June 2006 that concluded that the traditional forces of supply and demand could not fully account for increases in the prices of energy commodities.[31] Also, according to an energy firm, an investment bank, an academic, and hedge fund officials, increasing numbers of speculative traders in the market and rising trading volume placed upward pressure on futures prices. However, others, including investment bank and CFTC officials, have argued that speculators did not increase prices, but they provided liquidity and dampened volatility. Moreover, other investment banks, energy firms, and FERC officials told us that speculative trading in the futures markets can contribute to short-term price movements in the physical markets. However, they did not believe it was possible to sustain a speculative "bubble" over time because the two markets are linked and both respond to information regarding changes in supply and demand caused by such factors as the weather or geopolitical events. Therefore, in their view, speculation could not lead to artificially high or low prices over a long period.

Within the noncommercial trader category, the largest increases came from managed money traders—which generally trade for their own accounts rather than for others. Specifically, for crude oil, the average number of managed money traders that trade daily increased significantly from about 62 in July 2003 to about 128 in December 2006. At the same time, the number of smaller traders also grew significantly from an average of about 26 per day in July 2003 to an average of about 111 per day in December 2006. The number of managed money traders and smaller traders for unleaded gasoline, heating oil, and natural gas also increased similarly during that period. The number of commercial futures traders generally did not increase in a fashion similar to that of noncommercial traders.

As the number of traders has increased, so has the trading volume on NYMEX for all energy futures contracts, particularly crude oil and natural gas, as shown in figure 8. From 2001 through 2006, the average daily contract volume for crude oil increased by 90 percent and for natural gas increased by 93 percent. However, unleaded gasoline and heating oil experienced less dramatic growth in their trading volumes during this period.

Along with the strong growth of energy futures trading, the amount of energy derivatives traded outside of exchanges also appears to have increased significantly. However, comprehensive data on the trading volume of energy-related OTC derivatives are not available because OTC energy markets are not regulated. The Bank for International Settlements publishes data on worldwide OTC derivative trading volume for broader groupings of commodities that can be used as a rough proxy for trends in the trading volume of OTC energy derivatives.[32] According to these data, the notional amounts outstanding of OTC commodity derivatives—excluding precious metals, such as gold—grew by 854 percent from December 2001 through December 2005.[33] From December 2004 through December 2005, the notional amount outstanding increased by 214 percent to more than $3.2 trillion. Despite the lack of comprehensive energy-specific data on OTC derivatives, the recent experience of individual trading facilities revealed the growth of energy derivatives trading outside of futures exchanges. For example, according to an annual financial statement of the IntercontinentalExchange (ICE), the volume of contracts traded on ICE—including financially settled derivatives and physical contracts—increased by 438 percent, from more than 24 million contracts in 2003 to more than 130 million in 2006.

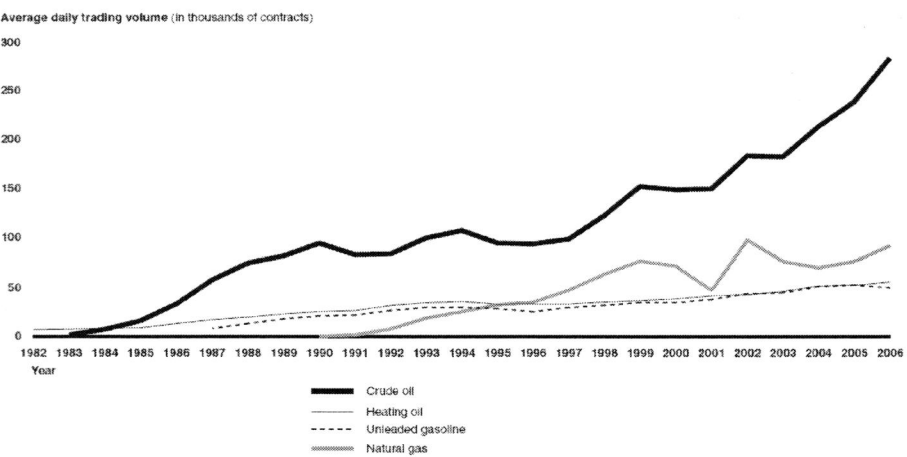

Source: GAO analysis of NYMEX data.

Note: The trading volume data for unleaded gasoline include the RB contract introduced on NYMEX in October 2005. The start dates for these commodities varied for these NYMEX contracts.

Figure 8. Average Daily Trading Volume for Crude Oil, Heating Oil, Unleaded Gasoline, and Natural Gas Futures Contracts, 1982–2006.

Several Factors Have Caused Changes in the Energy Markets... 29

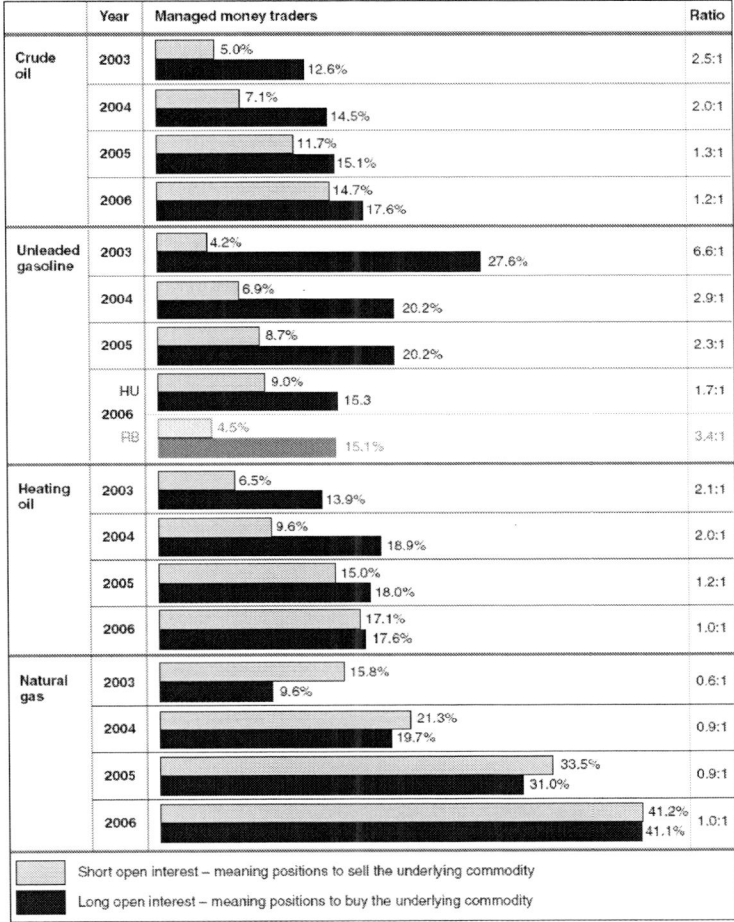

Source: GAO analysis of CFTC data.
Note: Data for 2003 were for July through December. The percentages indicate what portion of long and short open interest was held by managed money traders. For example, in 2004, managed money traders held 14.5 percent of the total long open interest for crude oil and 7.1 percent of the total short open interest. Because data are not included for all categories of traders, the percentages for these categories within a particular period do not total 100. These data should be viewed as a general overview of managed money traders' positions. They do not provide insights into how traders' individual positions changed over time. Our data for 2006 include contract trading data for RB and for the gasoline—HU— that began to be replaced by RB.

Figure 9. Percentage of Long and Short Open Interest in Futures and Options for Managed Money Traders, July 2003–December 2006.

While some market observers believed that managed money traders were exerting upward pressure on prices by predominantly buying futures contracts, CFTC data reveal that, from the middle of 2003 through the end of 2006, the trading activity of managed money participants became increasingly balanced between buying and selling. According to basic futures market theory, a trader speculating and holding an outstanding position to buy the commodity—a long open interest position—expects that the price of the commodity will rise, while a trader holding an outstanding position to sell the commodity—a short open interest position—expects that the price will decline. As shown in figure 9, according to CFTC data, from July 2003 through December 2003 managed money traders' ratio of long open interest in crude oil to short open interest was about 2.5:1, suggesting a strong expectation that prices would rise, on average, throughout that period, which they did. By 2006, this ratio fell to 1.2:1, suggesting that managed money traders as a whole were more evenly divided in their expectations about future prices. Managed money trading in unleaded gasoline, heating oil, and natural gas showed similar trends. Although for natural gas, open interest was more often short than long, suggesting a general expectation that prices would decline, which largely did not occur until 2006. Also, the relatively high percentage of open interest for natural gas held by these traders in 2006—surging to just over 40 percent—was perhaps due to the increased volatility of natural gas futures prices from 2005 to 2006, which provided traders with more opportunities for profit (or loss).

Chapter 5

CFTC OVERSEES EXCHANGES AND HAS LIMITED AUTHORITY OVER OTHER DERIVATIVES MARKETS

Energy products are traded on multiple markets, which are subject to varying levels of CFTC oversight and regulation. Under the CEA, CFTC regulatory oversight is focused on conducting the surveillance of futures exchanges, protecting the public, and ensuring market integrity. CFTC collects and analyzes trading position information on futures exchanges, which is central to this oversight. The information is subsequently published at highly aggregated levels in the commission's COT reports, and it helps to provide transparency to the market. However, these public reports have been criticized because the informational categories for traders do not accurately reflect energy market activity. While CFTC's oversight is focused on futures exchanges, the number of exempt commercial markets for trading energy commodities, which are not subject to general CFTC oversight, have grown. However, traders in these markets are subject to the CEA's antimanipulation and, where applicable, antifraud provisions.[34] Also, exempt commercial markets must provide CFTC with data for certain contracts and notify CFTC if cash markets use exempt market prices to price their transactions (although that has not occurred).[35] Energy products also are traded off exchange (referred to as OTC) and are not subject to direct CFTC oversight and regulation. However, as we have previously noted, certain types of off-exchange transactions are subject to antifraud and the antimanipulation provisions of the CEA, which CFTC has the authority to enforce. In addition, contract participants may be subject to other regulatory authority on the basis of their role in the physical market. To enhance its ability to detect and deter price manipulations, CFTC has published for comment a proposal to amend part 18 of its regulations to obtain from traders that have large (reportable) positions in an exchange-traded commodity information about their off-exchange positions in the same commodity.[36] CFTC also held a hearing in September 2007 to examine trading on regulated exchanges and exempt commercial markets, which included an assessment of price discovery and the implications for CFTC oversight.

CFTC HAS GENERAL OVERSIGHT AUTHORITY OVER FUTURES EXCHANGES, BUT ITS PUBLICLY REPORTED INFORMATION ON THESE EXCHANGES HAS NOT KEPT PACE WITH CHANGING MARKET CONDITIONS

Under the CEA, CFTC has general oversight authority over futures exchanges such as NYMEX. These exchanges receive CFTC approval to list futures and options contracts for trading and are subject to direct CFTC regulation and oversight. To be a regulated futures exchange, an exchange must demonstrate to CFTC that the exchange complies with (1) the criteria for designation under section 5(b) of the CEA for, among other things, the prevention of market manipulation, fair and equitable trading, the conduct of trading facilities, and the financial integrity of transactions conducted on the board; (2) the set of core principles under section 5(d) of the act establishing their regulatory responsibilities; and (3) the provisions on application procedures of part 38 of the CFTC rules.[37] According to CFTC officials, following procedures in the CEA, these exchanges may list new contracts, after certifying that they are in compliance with certain core principles, including ascertaining that the contracts are not readily susceptible to manipulation and monitoring trading to prevent price manipulation.[38]

CFTC's oversight is focused on fulfilling three strategic goals relating to futures exchanges. First, to ensure the economic vitality of the commodity futures and options markets, CFTC conducts its own direct market surveillance and also reviews on an oversight basis the surveillance efforts of these exchanges. According to CFTC officials, the commission monitors trading activity in futures markets and uses these trading data to analyze large positions that might be used to manipulate futures markets. In its oversight role, CFTC reviews new futures contracts to assess susceptibility to manipulation. To list a new futures contract, an exchange must file a written self-certification with CFTC and, if requested, must provide additional evidence, information, or data to CFTC on whether the contact satisfies CEA requirements and the commission's regulations or policies. Second, to protect market users and the public, CFTC has promoted sales practices and other customer protection rules applicable to futures commission merchants and other registered intermediaries.[39] In this connection, CFTC closely monitors the enforcement of registration and other requirements by the National Futures Association, which is an SRO responsible for regulating all firms and individuals conducting futures business with public customers. Third, to ensure the market's financial integrity, CFTC reviews the audit and financial surveillance activities of SROs. It also periodically reviews registered derivatives clearing organizations to ensure that they are effectively monitoring risks and protecting customer funds.

CFTC provides the public information on open interest in exchange-traded futures and options by commercial and noncommercial traders for various commodities in its weekly COT reports, which are relied upon by the public. Changing market conditions caused CFTC in 2006 to reassess COT reporting and its value to the public.[40] A trading entity generally gets classified as commercial by filing a statement with CFTC that it is commercially "engaged in business activities hedged by the use of the futures or option markets." To ensure that traders are classified with accuracy and consistency, commission staff review this self-classification and may reclassify a trader if staff have additional information about the trader's use of the markets. A trader may be classified as commercial in some commodities and as noncommercial in other commodities. A single trading entity cannot be classified as both commercial and noncommercial in the same commodity. Nonetheless, a multifunctional organization that has more than one trading entity may have each trading entity classified separately in a commodity. For example, a financial organization trading in financial futures may have a banking entity whose positions are classified as commercial and have a separate money-management entity whose positions are classified as noncommercial.

Recently, CFTC observed that the exchange-traded derivatives markets, as well as derivatives trading patterns and practices, have evolved, leading CFTC to question whether the commercial and noncommercial categories of today's COT reports appropriately classify trading practices. In June 2006, CFTC issued a notice in the *Federal Register* that it was undertaking a comprehensive review of the COT reporting program out of concern that the reports in their present form might not accurately reflect the commercial or noncommercial nature of positions held by nontraditional hedgers, such as swaps dealers.[41] On the basis of the comments received in response to the notice, in December 2006, CFTC announced the initiation of a 2-year pilot program for publishing a supplemental

COT report that would contain, in addition to categories for noncommercial and commercial positions, a category showing aggregate futures and options positions of index traders in 12 selected agricultural commodities. In explaining the program, CFTC observed that the "index traders" category would include traders that also were included in the noncommercial and commercial categories:

"In addition, the Commission will begin publishing a supplemental COT report that includes, in a separate category, the positions of commodity index traders in certain physical commodity futures markets. These so-called 'Index Traders' will be drawn from both the current Noncommercial and the Commercial categories. Coming from the Noncommercial category will be managed funds, pension funds and other institutional investors that generally seek exposure to commodity prices as an asset class in an unleveraged and passively managed manner using a standardized commodity index. Coming from the Commercial category will be entities whose positions predominantly reflect hedging of OTC transactions involving commodity indices—for example, swap dealers holding long futures positions to hedge short OTC commodity index exposure opposite institutional traders such as pension funds. These latter position holders are those traders described in the request for comments as 'non-traditional commercials.'"

CFTC stated that the pilot program for reporting of commodity index trading did not include energy and metals markets because the large trader data currently available to the commission would not permit an accurate breakout of index trading in these markets. According to CFTC, swap dealers, who use futures markets to hedge commodity index transactions in the OTC market, conduct most trading of commodity index-related futures. However, these swap dealers also may engage in OTC derivative transactions on energy or metals prices directly and conduct cash transactions in the underlying energy or metals markets. As a result of these activities, the overall futures positions held by swap dealers in energy and metal futures markets may not necessarily correspond closely with the hedging of OTC commodity index transactions. The commission stated that including these traders in the new index trader category would not enhance market transparency. Furthermore, it did not want to delay publication of the new COT report while it continues to study whether it is feasible to publish meaningful reports for other markets. The objective of the pilot program is to improve the transparency of an evolving market by separately reporting the positions of index traders. Similarly, the increasing volume of off-exchange trading in energy derivatives and the recent volatility of energy commodity prices justify considering whether a COT category of futures positions held by participants in off-exchange energy markets also could enhance transparency. CFTC said it will assess the relevance and usefulness of the new reporting and study whether it is possible and appropriate to expand the supplemental report to include data for other physical commodity futures markets.

Significant changes in the energy markets also may lead CFTC to further examine the usefulness, accuracy, and relevance of reported information to users. According to CFTC officials, energy trading has seen the entry of new market participants. For example, investment banks, hedge funds, and swaps dealers have become significant market participants. Moreover, according to industry analysts and representatives from investment banks and large oil companies, some commercial participants only hedge, some only speculate, and others both hedge and speculate in the energy markets. While some commercial participants may hedge and speculate in the same energy market, CFTC classifies these entities as commercial participants. CFTC has not been able to identify new categories for traders of energy commodities. Such reporting can distort the accuracy and relevance of reported information to users and the public, thereby limiting the usefulness of the information reported to the public as well as information used by traders.

CFTC AUTHORITY OVER EXEMPT COMMERCIAL MARKETS CONSISTS OF ENFORCING THE ANTIFRAUD AND ANTIMANIPULATION PROVISIONS OF THE CEA AND ADMINISTERING CERTAIN REPORTING REQUIREMENTS

In contrast to the direct oversight provided to futures exchange, exempt commercial markets are not subject to CFTC's general oversight authority. According to CFTC officials, as these markets have grown in prominence, some market observers have questioned their role in the energy markets. Trading energy derivatives on exempt commercial markets is permissible only for eligible commercial entities. While not subject to general CFTC oversight, these markets are subject to CFTC rule 36.3, which provides for the dissemination of exempt commercial market trading data should exempt commercial market prices be used to price cash markets and contains notification, recordkeeping, and reporting requirements.[42] Also, exempt commercial market participants are subject to CFTC's enforcement authority for the antimanipulation and antifraud provisions of the CEA.[43] These markets are not required to register with CFTC, but must notify CFTC that they are operating as an exempt commercial market and comply with certain CFTC informational, recordkeeping, and other requirements.[44]

Specifically, CFTC promulgated rule 36.3 under two subsections of the CEA. One subsection authorizes CFTC to prescribe rules if necessary to ensure the timely dissemination of price, trading volume, and other trading data for a derivative traded on an exempt commercial market if the commission determines that the electronic trading facility used by the market performs a significant price discovery function for transactions in the cash market for the commodity underlying the derivative.[45] The other subsection establishes notification, recordkeeping, and reporting requirements for exempt commercial markets.[46] The rule requires, among other things, that the electronic trading facility in an exempt commercial market must notify CFTC of its reliance on the exemption and provide CFTC with price, quantity, and other data on contracts that average five or more trades a day over the most recent quarter for which they are relying on the CEA exemption. The facility also must maintain a record of allegations or complaints they receive concerning instances of suspected fraud or manipulation and provide CFTC with a copy of the record. CFTC officials said that the reports include transaction-level data, such as quantity and price, for all trades in products meeting the criteria, but not the identities of counterparties to the trades. These officials said that three exempt commercial markets—ICE, the Natural Gas Exchange, and ICAP—currently provide the rule 36.3 trade information reports; in the past, the Optionable and ChemConnect exempt commercial markets also provided these reports. For example, ICE officials told us that for their OTC activities they keep records for all of the products traded on their platform, and report to CFTC on liquid markets (those averaging five trades a day) and any complaints received from market participants. ICE officials said that CFTC often asks ICE for detailed information about participants that are putting up bids and offers and about all of the trades executed in a day.

CFTC officials said that the other electronic exchanges have provided notice that they are operating in reliance on the CEA exemption, but they have not provided rule 36.3 trade information reports. CFTC officials explained that an electronic exchange only has to provide information reports if it meets the threshold for reporting, which includes averaging five trades per day in the relevant contract. These officials also said that they do not actively check to determine whether the thresholds are being met.

To date, no exempt commercial market or CFTC has determined if cash markets for energy commodities routinely use exempt market prices to price their transactions. According to CFTC officials, an exempt commercial market or CFTC may determine, using certain criteria, if the market serves such a price discovery function. Exempt markets that serve such a function become subject to certain public reporting requirements. According to CFTC officials, the commission has not made such a determination for two reasons. First, they said that the only consequence of serving a price discovery function under current law is that the exempt commercial market must publish its prices. They noted that this is a circular argument because it is the public availability of pricing information that enables the exempt commercial market to serve a price discovery function. Second, they said that this is a low priority. In their view, the current fiscal situation does not allow CFTC to send its economists into the field on matters such as this that would not go before the commission. Also in their view, even if the markets served a price discovery function, no significant consequence would entail because of the circularity argument. However, in light of the growth of trading on ICE and the lessons learned from the Amaranth crisis, CFTC held a hearing in September 2007 to examine trading on regulated exchanges and exempt commercial markets.[47] The hearings included an assessment of price discovery in these markets and the implications for CFTC oversight of these markets.

Since 2001, 17 facilities have notified CFTC that they had begun operating as exempt commercial markets (see table 1). According to CFTC officials, 11 of these markets currently offer, or had offered, transactions in energy products, with 8 now operational. Some of these markets have become important players in the trading of energy products. ICE, in addition to the exempt swap contracts it trades in its capacity as an exempt commercial market, is the trading platform for physical commodities, including spot and forward contracts, which routinely involve delivery. According to CFTC officials, some in the industry assert that ICE is the trading platform for an estimated 70 percent of the spot trading for natural gas.[48] Another exempt commercial market, ChemConnect, advertises that data and news providers, such as Bloomberg and Dow Jones Energy Services, rely on it to provide accurate, timely information on energy products. Furthermore, the Web site for the HoustonStreet Exchange indicates that it serves as an electronic trading facility for crude oil and refined products also traded on NYMEX. While there has been significant growth in the number of electronic exchanges, CFTC officials said that they receive trade information reports from only 2—ICE and the Natural Gas Exchange. According to CFTC officials, they have no evidence that the others meet the minimum threshold trading volume for reporting.

Table 1. Exempt Commercial Markets, Dates They Filed Notice with CFTC to Operate as an Exempt Commercial Market, and Commodities Traded on Each Market, 2001–2006

Notification date	Exempt commercial market	Commodity category
2006	ChemConnect	Energy products
2003	Chicago Climate Exchange	Emission allowances
2002	Commodities Derivative Exchange	Metals
2002	HoustonStreet Exchange	Energy products
2006	ICAP Commodity and Commodity Derivatives Trading System	Energy products
2006	ICAP Electronic Trading Community	Natural gas and its derivatives
2005	ICAP Hyde Limited Trading System	Forward freight agreements
2001	IntercontinentalExchange	Precious metals, base metals, and energy products
2001	International Maritime Exchange	Freight rates
2002	Natural Gas Exchange	Energy products
2006	NetThruPut	Condensates and liquefied petroleum gas
2001	Optionable	Energy products
2003	Spectron Live.com Limited	Liquefied petroleum gas
2003	TFS Energy	Weather derivatives
2005	Trade Capture	Energy products
2002	Tradespark	Energy products, weather indexes, and emission allowances
2003	Tradition Financial Services Pulp and Paper Division	Pulp and paper products

Source: GAO analysis of CFTC data.

ALTHOUGH CFTC CAN ENFORCE ANTIMANIPULATION AND APPLICABLE ANTIFRAUD PROVISIONS OF THE CEA IN OTC ENERGY DERIVATIVES MARKETS AND EXEMPT COMMERCIAL MARKETS, VIEWS VARY ABOUT THE SUFFICIENCY OF ITS REGULATORY AUTHORITY WITH RESPECT TO OFF-EXCHANGE ENERGY DERIVATIVES

Energy derivatives also may be traded OTC, under the conditions and restrictions in the CEA for exempt commodities. The act exempts from most of its provisions transactions in exempt commodities into which large market participants enter and that are not traded on a trading facility. In addition, the act exempts from most of its provisions transactions in exempt commodities traded on an electronic trading facility, as long as large commercial traders (defined in the act as "eligible commercial entities") enter into them on a principal-to-principal basis.[49] Bilateral OTC derivatives contacts are viewed as private transactions between sophisticated counterparties, and there is no requirement for parties involved in OTC transactions to disclose details of their transactions. Because OTC derivatives are contractual agreements, each party is subject to and assumes the risk of nonperformance by its counterparty. This is different from exchange-traded derivatives, where a central clearinghouse stands behind every trade. Thus, according to officials of the International Swaps and Derivatives Association, in the OTC context it is vitally important that one has confidence in the creditworthiness and trustworthiness of one's counterparty. While these markets generally are not subject to direct CFTC oversight, CFTC has the authority to enforce antifraud and antimanipulation provisions of the CEA in connection with transactions in exempt commodities that take place through an electronic trading facility, and that are entered bilaterally without being subject to negotiation.[50] Several of the enforcement actions filed by CFTC since 2001 addressed the use of false reporting in an attempt to manipulate energy prices on NYMEX.

In addition to being subject to certain provisions of the CEA, the participants in these contracts may be subject to other regulatory authorities on the basis of their activities in the physical market. For example, certain actions—such as the buying and selling of a physical energy commodity by traders, such as hedge funds—may fall under the regulatory authority of FERC, which regulates the interstate transmission of physical commodities, such as natural gas, oil, and electricity, to protect energy consumers. Also, certain OTC derivative activities conducted by commercial banks are subject to oversight by the appropriate bank regulator. For example, commercial banks that engage in OTC derivatives are overseen by their relevant regulator, such as the Office of Comptroller of the Currency or the Board of Governors of the Federal Reserve System with respect to how their derivatives trading satisfies requirements of the banking laws. Likewise, SEC also has oversight authority over investment banks' activities that fall under its regulatory purview. These regulators do not regulate the specific transactions or maintain oversight of OTC derivatives as a class of instruments or markets; they regulate the entities that enter into the contracts or that act as dealers, counterparties, or both.

While some observers have called for more oversight of OTC derivatives, most notably for CFTC to be given greater oversight authority of this market, others oppose any such action as unnecessary. Supporters of more CFTC oversight authority believe that regulating OTC derivatives markets is needed to protect the regulated markets and protect consumers from potential abuse and possible manipulation. One of their concerns is that because there is little information available about the size of this market or the terms of the contracts, CFTC may not be assured that trading on the OTC market is not adversely affecting the regulated markets and, ultimately, consumers. Specifically, some have mentioned that, unlike trading on a regulated exchange, OTC derivatives are not subject to any routine reporting requirements. Some have suggested that a combination of quantitative and qualitative information (such as whether derivatives are used mainly for trading or hedging purposes, and notional amounts by derivatives category) be collected.[51]

However others, including the President's Working Group, have concluded that OTC derivatives generally are not subject to manipulation because contracts are settled in cash on the basis of a rate or price determined in a separate, highly liquid market and these OTC transactions do not serve a significant price discovery function.[52] The Working Group also noted that if electronic markets were to develop and serve a price discovery function, then consideration should be given to enacting a limited regulatory regime aimed at enhancing market transparency and efficiency through CFTC, as the regulator of exchange-traded derivatives.

However, because of the lack of reported data about this market, addressing concerns about its function and effect on regulated markets and entities would be a challenge. CFTC officials have said that they have reason to believe these off-exchange activities affect prices determined on a regulated exchange. In a June 2007 *Federal Register* release clarifying its large trader reporting authority, CFTC noted that having data about the off-exchange positions of traders with large positions on regulated futures exchanges could enhance the commission's ability to deter and prevent price manipulation or any other disruptions to the integrity of the regulated futures markets.[53] According to CFTC officials, the commission also has proposed amendments to clarify its authority under the CEA to collect information and bring fraud actions in principal-to-principal transactions in these markets, thus enhancing CFTC's ability to enforce antifraud provisions of the CEA.[54]

Also, in August 2007, CFTC announced plans to conduct a hearing to begin examining more closely the trading on regulated exchanges and exempt commercial markets. The September 2007 hearing focused on a number of issues, including

- the current tiered regulatory approach established by the CFMA and whether this model is beneficial;

- the similarities and differences between exempt commercial markets and regulated exchanges, and the associated regulatory risks of each market; and
- the types of regulatory or legislative changes that might be appropriate to address any identified risks.

Chapter 6

CFTC ENGAGES IN SURVEILLANCE ACTIVITIES AND ENFORCEMENT ACTIVITIES, BUT THE EFFECTIVENESS OF THESE ACTIVITIES IS LARGELY UNCERTAIN

CFTC provides oversight for commodity futures markets through routine surveillance, analysis of market data, and inquiries of market participants and others. The commission uses information gathered from surveillance activities to identify unusual trading activity and possible market abuse. In particular, CFTC's LTRS provides essential information for surveillance, and LTRS provides information on the majority of all trading activity on futures exchanges. CFTC staff also rely on data from other sources and on their experience to identify potential problems, reporting unresolved potential market problems to the commission. NYMEX also conducts its own surveillance activities. According to CFTC and industry officials, CFTC and NYMEX contact traders to collect additional information about questionable trading practices. CFTC staff also said that they routinely investigate traders with large open positions, but the staff added that they do not routinely maintain information about such inquiries, thereby making it difficult to determine the usefulness and extent of these activities. In addition, CFTC coordinates its surveillance activities with other federal, state, and foreign authorities. While CFTC's surveillance authority is limited to futures exchanges, the commission's enforcement authority for manipulation and fraud extends to both exchange-based trading and off-exchange trading in exempt commodities, such as energy products. According to data provided by CFTC, in recent years, it has used its enforcement authority to file enforcement actions for almost 300 cases, more than 30 of which involved energy-related commodities. However, as with programs operating in regulatory environments where performance is not easily measurable, evaluating the effectiveness of CFTC's enforcement activities is challenging because of the lack of effective outcome-based performance measures. CFTC's enforcement program received mixed ratings in a recent OMB review because CFTC could not fully demonstrate the effectiveness of its enforcement activities.

CFTC OVERSIGHT INCLUDES SURVEILLANCE OF ENERGY FUTURES TRADING, BUT THE FULL EXTENT OF FOLLOW-UP ACTIVITIES IS UNCERTAIN

CFTC conducts regular market surveillance and oversight of energy trading on NYMEX and other futures exchanges. These activities include focusing on detecting and preventing disruptive practices before they occur and keeping the CFTC commissioners informed of possible manipulation or abuse. In addition to conducting direct surveillance of trading in energy futures markets on NYMEX, CFTC focuses on NYMEX's compliance with appropriate CEA core principles, including monitoring of trading to prevent price manipulation and enforcing position limits and position accountability rules. In conducting its own surveillance activities, NYMEX may bring enforcement actions when violations are found. CFTC staff also investigate traders with large open positions and document cases of improper trading.

CFTC Oversees Trading on Futures Exchanges

According to CFTC officials, CFTC staff at three regional offices provide much of the market oversight and monitor daily trading activity. For instance, CFTC's New York Regional Office employs seven economists, who look for unusual trading and potential market manipulations in all futures contracts traded on New York futures exchanges. The New York regional staff obtain information from both market participants and NYMEX to monitor energy trading activity. New York CFTC staff stated that each morning, about 160 firms electronically submit large trader position data from the previous day to CFTC. CFTC headquarters receives these data and makes them available on a network to its field offices. Staff review these data for potential errors or omissions and then populate the LTRS, a database that staff use in conducting their surveillance activities.

CFTC staff also said that they rely on the commission's integrated surveillance system (ISS), which contains surveillance data that CFTC has collected from the futures exchanges, clearing members, foreign brokers, and large traders. According to CFTC's 2005 performance and accountability report, ISS is a critical application to support futures and options data market surveillance.[55] This system provides continuously updated trading data on holders of large futures and option positions that CFTC staff uses daily to monitor futures and option trading, detect potential problems, and identify trends in the marketplace. According to CFTC officials, ISS also is used to facilitate analysis of data received from exempt markets as a result of special calls for information. For example, pursuant to separate special calls issued in April, September, and December, 2006, ICE now continuously provides the commission with large trader position data. The commission also issued enforcement-related special calls seeking data for two individual ICE market participants in September 2006 and February 2007.

The LTRS, which is part of ISS, is a comprehensive system for collecting information on market participants, a key information source for CFTC's market surveillance program and essential for monitoring markets and identifying and resolving potential problems involving market congestion, manipulation, and speculative position limits.[56] Congestion may occur when traders holding short positions are attempting to cover their positions but are unable to find an adequate supply of contracts provided by traders with long positions or by new sellers willing to enter the market, except at sharply higher prices. In conducting their daily surveillance activities, CFTC officials said they analyze the trading data for indications that individual traders may be attempting to manipulate the market. This activity involves (1) looking for traders having unusually large market positions relative to open interest—the total number of futures contracts that have been entered into and not yet liquidated by an offsetting transaction or fulfilled by delivery—and deliverable supply and (2) examining the potential for disruption at expiration and sharp moves in the market. If certain positions pose concerns to CFTC staff, they can request additional information from a reporting firm or trader about trading and delivery activity.

CFTC staff also analyze trading using data from other sources. CFTC officials said that the staff look at price movements and price relationships—especially in the natural gas, crude oil, heating oil, and unleaded gasoline markets—using commercial information sources, such as Bloomberg, Gas Daily, Reuters, and other market sources. They also obtain information about traders by monitoring their Web sites and use NYMEX's and EIA's Web sites and Lexis-Nexis, as well as firms' Web sites. CFTC staff said that they are in regular contact with exchange officials, who have data on clearing members and trading activity. They also obtain surveillance information from other units within the commission and from tips by the public.

While CFTC data and other market collections are focused on identifying potential market disruptions and manipulations, staff also rely on their experience to identify potential problems. According to CFTC staff, the New York Regional Office staff assigned to surveillance of energy trading have many years of experience, either doing surveillance work for CFTC or in the futures industry, in general. Experienced staff are needed because, according to CFTC staff, analyzing market data is an art as well as a science. CFTC staff referred to the traditional test for manipulation set forth in the commission's *Indiana Farm* decision as a commonly recognized statement of the elements that are necessary to prove manipulation.[57]

According to CFTC staff, when a potential market problem has been identified, surveillance staff generally contact the exchange or traders to gather additional information. They said that surveillance staff may ask exchange employees, brokers, or traders questions to confirm positions and determine the intent of traders. They added that staff may express concern about the size of positions or possible actions by traders and caution traders to act responsibly.[58] According to the staff, CFTC's Division of Market Oversight may issue a warning letter or make a referral to the Division of Enforcement to conduct a nonpublic investigation into the trading activity. Markets where surveillance problems have not been resolved may be included in reports presented to the commission at weekly surveillance meetings. These reports provide information on traders with the four largest long and short positions; other market information, including delivery information; and background on the contact. According to CFTC staff, CFTC commissioners review the reports; discuss the situations with surveillance staff; and, if appropriate, consider other possible remedial actions, such as suggesting that the exchange take emergency action. If necessary, the commission itself may take emergency action.

If these actions do not resolve the issue or if an exchange fails to resolve a problem by taking actions that the commission deems appropriate, CFTC can order an exchange to take emergency actions. These actions include limiting trading, imposing or reducing limits on positions, requiring the liquidation of positions, extending a delivery period, or suspending trading. The commission has taken such emergency actions four times in its history, but never for energy markets.

CFTC Engages in Surveillance Activities and Enforcement Activities... 47

In addition to CFTC's surveillance of NYMEX and trading on the exchange, NYMEX conducts its own surveillance activities and, if violations are found, brings its own enforcement actions. NYMEX is responsible for enforcing its own standards and CFTC's standards embodied in its rules governing the exchange, and its surveillance program is designed to monitor for possible manipulation by market participants. If NYMEX staff find potential violations, they will gather information and, if needed, take enforcement actions. For example, according to officials at a large refiner, NYMEX staff call them nearly every month about a large trade to make sure that their physical (or wet) barrels have moved and that their trade is not a price-setting mechanism or market ploy. Refiner officials added that even though NYMEX staff know they are a big refiner, they will examine their trades to see the actual signed contract to make sure it is valid. In their view, NYMEX staff are vigilant, as they should be. Officials from a hedge fund also said that both NYMEX and CFTC staff monitor their positions carefully and, as a speculator, would be notified immediately by NYMEX and CFTC if they were over the trading limits on any day. When asked about what weaknesses in the structure, monitoring, or enforcement mechanisms of derivative markets might allow for market manipulation, one market observer responded that he was not aware of any such weaknesses. Appendix III contains detailed discussion of NYMEX surveillance activities and enforcement actions.

Actions Taken by CFTC Staff to Inquire About Potential Problems May Not Always Be Documented

CFTC staff routinely make inquiries about traders with large open positions approaching expiration, but formal records of their findings are only kept in cases where there is evidence of improper trading. If LTRS data reveal that a trader has a large open market position that could disrupt markets if it were not closed before expiration, CFTC staff would contact the trader to determine why the trader had the position and what plans the trader had to close the position before expiration or to ensure that the trader was able to take delivery. If the traders provided a reasonable explanation for the position and a reasonable delivery or liquidation strategy, staff said that no further action would be required. CFTC staff said they would document such contacts on the basis of their importance in either informal notes, e-mails to supervisors, or informal memorandums. No formal record would be made of the inquiry, according to one CFTC official, unless there was a signal indicating improper trading activity. Without such data, CFTC's measures of the effectiveness of its actions to combat fraud and manipulation in the markets will not reflect this surveillance activity, and CFTC management might miss opportunities to both identify trends in activities or markets and better target its limited resources.

CFTC staff added that all surveillance projects and activities that require a minimum number of hours of work are tracked by quarterly statistical reports, including those futures expirations with large trader or deliverable supply problems. They said that expirations are routinely monitored by economists and reviewed with their supervisors through weekly surveillance reports. Economists are responsible for the analytical review of cash and futures market developments, including the assessment of supply and demand factors, basis and spread relationships, the adequacy of deliverable supply, large trader positions and position changes, large trader histories, and the potential for group trader activity. CFTC staff said that their economists keep their supervisors and the commission informed of potential problems as they arise.

CFTC COORDINATES ITS SURVEILLANCE AND OVERSIGHT ACTIVITIES WITH OTHERS

In addition to keeping CFTC commissioners apprised of surveillance activities and specific cases that may require action, CFTC coordinates its surveillance and oversight activities with other federal agencies, states' attorneys general, and foreign regulators. CFTC officials told us that through the Division of Enforcement's Office of Cooperative Enforcement, which was created in 2002, they conduct outreach efforts to other financial regulators at the federal and state levels. Specifically, CFTC and FERC coordinate oversight and enforcement activities and have a memorandum of understanding that provides for the exchange of data. FERC regulates the interstate transmission of natural gas, oil, and electricity, and it audits natural gas sellers' compliance with the protocols outlined by FERC for reporting sales to index publishers like Platts, a company that compiles information on oil, natural gas, and electricity and other energy commodities and provides industry reports on commodity prices. If futures transactions are thought to affect transactions within FERC's jurisdiction, then FERC and CFTC may coordinate their oversight and enforcement work by sharing data as provided in the memorandum. In pursuing potential market abuse cases, such as individuals trying to manipulate energy spot prices to benefit their futures market positions, FERC officials said that FERC will tend to take the lead when abuses occur in the physical markets. FERC officials also said that CFTC will tend to take the lead when abuses occur in the futures markets. In July 2007, FERC filed two market manipulation cases that, according to a commission announcement, was the first time the agency used its enforcement authority under the Energy Policy Act of 2005 and its former market manipulation rule. According to CFTC officials, CFTC has filed 38 cases over the past 6 years that have focused on conduct in both the cash and futures markets (see app. IV). CFTC and FERC also may work with DOJ on certain cases.

In addition, CFTC officials said that, on occasion and when warranted by the circumstances, CFTC has shared large trader information with certain agencies, such as the Department of the Treasury, the Board of Governors of the Federal Reserve System, and the Federal Reserve Bank of New York, to address issues of common concern to the agencies. For example, in the aftermath of the financial difficulties in 1998 of Long Term Capital Management, a large hedge fund, CFTC shared information on the hedge fund's exchange trading activity with members of the President's Working Group. Because coordinating requires judgments about what information would need to be and could be shared and about how best to share it, we concluded in a 1999 report that the regulators are in the best position to determine the most effective ways to enhance their coordination.[59] CFTC also shares information with other members of the President's Corporate Fraud Task Force at their quarterly meetings on antifraud cases.

CFTC ENERGY-RELATED ENFORCEMENT ACTIONS GENERALLY INVOLVED FALSE REPORTING AND ATTEMPTED MANIPULATION, AND ENFORCEMENT ACTIONS OFTEN ARE COORDINATED WITH OTHER AUTHORITIES

CFTC's Division of Enforcement is charged with enforcing the antimanipulation sections of the CEA, including sections 6(c), 6(d), and 9(a)(2). In particular, section 9(a)(2) sets forth the commission's antimanipulation and false reporting authority in cash and futures markets.[60] In determining whether violative conduct has occurred, CFTC officials told us that the Division of Enforcement has broad investigatory authority to obtain records and testimony, including subpoena authority, under a commission order. They added that upon conclusion of an investigation, which is routinely nonpublic, the division may recommend enforcement action if warranted.

The enforcement actions CFTC has taken in its energy-related cases generally have involved false public reporting as a method of attempting to manipulate prices on both the NYMEX futures market and the off-exchange markets. CFTC officials said that from October 2000 to September 2005, the commission initiated 287 enforcement cases and more than 30 of these cases involved energy trading, including actions against Enron and others. For example, according to CFTC data, from 2001 through 2005, CFTC levied fines totaling $305 million in actions alleging attempted manipulation of the price of natural gas (see app. IV for more detailed information). Most of these cases charged attempted manipulation by means of falsely reporting natural gas trading information to energy index firms, such as Platts, that calculate surveys or indexes of natural gas prices for various physical delivery points (hubs) throughout the United States. Generally, these cases involved allegations of various defendants knowingly disseminating false information in an effort to skew the indexes for their financial benefit or for other reasons. Participants in the natural gas markets use the indexes for price discovery and assessing price risk. Many of the actions were initiated on the basis of information that came from sources other than CFTC surveillance activities, or those of NYMEX, because they involved activities outside of NYMEX. As one major oil company official told us, in his view, CFTC and FERC vigorously pursued attempts by traders to manipulate the market.

Most recently, on August 1, 2007, the commission entered an order imposing a $1 million penalty against Marathon Petroleum Company, LLC, for attempting to manipulate spot cash crude prices by attempting to influence the Platts market assessment. On July 25, 2007, the commission commenced an action against Amaranth and others for attempted manipulation of NYMEX natural gas futures prices. Also on July 26, 2007, the commission commenced an enforcement action on Energy Transfer Partners, L.P., and others for attempted manipulation of physical natural gas prices.

Regarding energy futures, CFTC coordinates its enforcement activities with NYMEX officials and various other federal, state, and foreign authorities. CFTC staff stated that they meet periodically with NYMEX Compliance Department officials to discuss enforcement activities, as appropriate, and have formal quarterly meetings to discuss mutual involvement in specific cases, including energy products. In addition to coordinating energy enforcement matters with NYMEX, as a regulator of derivatives trading, CFTC often will work with the regulator of the underlying commodity or affected market, whether the Department of Agriculture, FERC, or Treasury. CFTC does not have criminal authority but often works with DOJ on those cases involving violations of the CEA that DOJ believes warrant criminal prosecution.[61] DOJ officials stated that their focus has been on natural gas cases, which began with cases involving Enron. According to DOJ officials, their role complemented the regulatory roles of FERC and CFTC, and they have an effective working relationship with CFTC in terms of sharing case information. For example, pursuant to a memorandum of understanding with CFTC, in May 2006, FERC obtained information about trading in natural gas futures contracts that FERC used in support of an enforcement action against Amaranth that was initiated in July 2007.[62] On July 25, 2007, CFTC filed an action in the United States District Court for the Southern District of New York against Amaranth Advisors, L.L.C., Amaranth Advisors (Calgary) ULC, and Brian Hunter alleging, among other things, that the defendants intentionally and unlawfully attempted to manipulate the price of natural gas futures contracts on NYMEX on February 24, 2006, and April 26, 2006.

In another case, on June 28, 2006, CFTC brought an enforcement action against BP Products North America, Inc., alleging, among other things, that BP cornered the physical propane market and manipulated the price of propane in February 2004.[63] Also on June 28, 2006, DOJ announced that a former BP trader had pled guilty to conspiracy to manipulate and corner the physical propane market. FTC also has exercised its authority in the energy arena. Since 1980, FTC's focus in energy has been in reviewing mergers and acquisitions for anticompetitive behavior and investigating instances of possible collusion, price fixing, and other anticompetitive conduct. However, FTC staff told us that they generally did not coordinate their work with CFTC, but added that they would turn over any evidence of futures manipulation to CFTC. CFTC staff said that, as appropriate, CFTC also coordinates its antifraud enforcement activities with states' attorneys general, who often will assist in a case by acting as a co-plaintiff with CFTC. In turn, CFTC may detail an attorney to a state. CFTC staff said that they also may work with international authorities, such as the United Kingdom's Financial Services Authority, on cases involving activities in more than one nation.

CFTC'S ENFORCEMENT PROGRAM RECEIVED A MIXED OMB RATING BUT LACKS EFFECTIVE OUTCOME-BASED PERFORMANCE MEASURES

Although CFTC has undertaken enforcement actions and levied fines, OMB's most recent 2004 Program Assessment Rating Tool (PART) assessment of the CFTC enforcement program was mixed. OMB designed PART to provide a consistent approach to assessing federal programs in the executive budget formulation process. PART is a standard series of questions meant to serve as a diagnostic performance tool, drawing on available program performance and evaluation information to form conclusions about program benefits and recommend adjustments that may improve results. In the assessment, OMB rated the enforcement program as "Results Not Demonstrated" and said that the enforcement program lacked performance measures that illustrate whether the program meets its overall objective. However, CFTC's existing performance measures show that it brings substantive cases in a timely manner and "is well designed to meet its objectives [of protecting commodity futures and options market users and the public from fraud, manipulation, and abusive practices related to the sale of certain commodities through the enforcement of laws against such practices] and to maximize the use of its resources." According to the PART assessment, the enforcement program has a clear purpose, addresses the public interest by ensuring adherence to the CEA and CFTC's regulations, is not duplicative of other government programs, is free of major design flaws, and is effectively targeted so that the resources address the program's purposes. OMB scored CFTC at 100 percent for the dimensions of both program purpose and design and program management, 71 percent for planning, and 67 percent for results and accountability. Compared with the other 96 programs that OMB identified as similar to CFTC's program, the comparable programs have much lower average scores for the dimensions of purpose and design (82 percent), program management (84 percent), and results and accountability (50 percent) and have a similar score for planning (73 percent).

CFTC's score of 71 percent for the planning dimension reflected OMB's assessment that CFTC included performance measures in its annual reports; used the actual results it achieved during the preceding fiscal year as a baseline for all of its performance measures and strove to set ambitious targets for its performance; was scrutinized on a regular basis by CFTC's Office of the Inspector General; had budget requests that were explicitly tied to accomplishment of the annual and long-term performance goals, and resource needs that were presented completely and transparently in the program's budget; and had taken meaningful steps to correct its strategic planning deficiencies. However, OMB also concluded that regarding the strategic planning dimension, the program had a limited number of long-term performance outcome measures that did not fully reflect the program's goals, and that the long-term measures and targets did not fully reflect the program's purposes. These measures included

- the percentage growth in market volume,
- the increase in the numbers of exchanges and clearinghouses,

- the percentages of SROs and clearing organizations that complied with the requirement to enforce their rules, and
- the percentage decreases in both the number of customers who lost funds because of alleged wrongdoing and the amount of funds that these customers lost.

CFTC enforcement staff stated that they face challenges in establishing measures to determine whether the enforcement program achieves its goal of deterring people from engaging in market manipulation or other abusive behavior.

According to OMB, CFTC's score of 67 percent on the program results and accountability dimension reflected its assessment that CFTC's enforcement program had demonstrated both (1) improved time efficiencies and cost-effectiveness in achieving its program goals and (2) our several evaluations of CFTC indicating that it was effective and achieving results.[64] OMB also reported that for fiscal year 2004, the enforcement program met all of its outcome measures and came close to meeting all of its output measures, with one exception. OMB further stated that the outcome-related measures established for enforcement do not fully reflect progress on meeting the program's overall goals.

While CFTC satisfied most but not all of OMB's PART criteria, it has fallen short in its ability to develop long-term performance outcome measures that are reflective of its program's goals and purposes. As OMB identified, CFTC has substituted proxy measures for outcome measures: that is, using measures such as percentage growth in market volume and increase in the number of exchanges and clearinghouses as proxies for protecting market integrity, and percentage decreases in both the number of customers who lost funds because of alleged wrongdoing as proxies for both protecting market integrity and consumers. We have found that managers in a regulatory environment where programs and activities are not easily measurable, as is the case with CFTC enforcement, have reported that it is particularly challenging to measure outcome-oriented performance and collect useful data.[65] However, there are a number of other ways to evaluate program effectiveness, such as using expert panel reviews, customer service surveys, and process and outcome evaluations. We have found with other programs that the form of the evaluations reflect differences in program structure and anticipated outcomes, and that the evaluations are designed around the programs and what they aim to achieve.[66] Without utilizing these or other methods to evaluate program effectiveness, CFTC is unable to demonstrate whether its enforcement program is meeting its overall objectives.

Chapter 7

CONCLUSION

The rise in energy prices can be and has been attributed to a variety of factors. From January 2002 through June 2006, the physical and derivatives markets both underwent substantial change and evolution. The physical energy markets experienced tight supply and increasing global demand, ongoing political instability in oil-producing regions, and other supply disruptions, which affected the prices of energy products. At the same time, increasing numbers of and different types of market participants were trading futures in search of higher returns, thereby increasing contract volume. Substantial growth in the exempt commercial and OTC markets also occurred. Determining the impact of any one factor is complicated because price changes in the physical and futures markets are closely linked and in the long run are influenced by the same market fundamentals. Generally, futures prices reflect traders' views of the impact of changes in the physical markets and spot prices are affected by these expressed views and vice versa. Given this interrelationship, it is not surprising that some market observers point to the changes in the energy futures and other derivatives markets as a possible explanation for price increases, while others, primarily the regulators, look to changes in the physical markets to explain the increases. However, given the changes in both markets, attributing causality to any one factor—much less a particular type of trading activity—is difficult. Regardless of the reason for the increases in prices, ongoing monitoring of both markets is warranted to ensure that the public interest is being protected as well as the integrity of the markets.

Related to concerns about rising prices, some market observers and others have questioned whether CFTC's authority is broad enough to protect investors from fraudulent, manipulative, and abusive practices. The scope of CFTC's authority varies, depending on the market where the commodity is traded. Some markets are available for retail trading and receive direct CFTC oversight, while others are limited to professional traders (such as OTC energy derivatives markets) and receive less oversight. Other markets are largely unregulated. Given the changes in these markets in general and the growth in off-exchange trading in particular as well as ongoing questions about the relationship between exchange-traded and off-exchange markets, a reexamination of the scope of CFTC's authority is warranted. The results of CFTC's hearings on its existing regulatory structure and the similarities and differences between exchange-traded and exempt markets may be instructive for such a reexamination. While participants on all sides of this issue have perspectives that call for further consideration, these are public policy decisions that ultimately will be made by Congress. Unless resolved, questions will continue about the scope of CFTC's authority.

In the interim, we have identified a number of process issues that CFTC can address to strengthen its enforcement and surveillance programs.

- First, CFTC has attempted to provide the public with more meaningful information through the COT reports. While this effort has expanded the reporting for some agricultural commodities, it has remained virtually unchanged for energy commodities that have a high level of public and industry interest. Not having complete information on trading in energy commodities impairs the ability of traders to make fully informed decisions.
- Second, CFTC's oversight of regulated exchanges involves a range of surveillance activities that have resulted in a number of commission-related enforcement actions. However, CFTC does not maintain complete records of its surveillance activities. Currently, the commission does not maintain written records on all surveillance follow-up activities, particularly in instances where no potential violation was found. Without such records, CFTC staff cannot fully demonstrate the actions they are taking to combat fraud and manipulation in the markets.
- Third, as is the case with most enforcement agencies, CFTC has had limited success in identifying meaningful outcome-based performance measures. However, agencies can use a variety of methods to evaluate program effectiveness, such as expert panel reviews, customer service surveys, and process and outcome evaluations. Without meaningful measures for program effectiveness, CFTC may be missing opportunities to identify significant trends in certain activities or markets and to better target its limited resources.

MATTER FOR CONGRESSIONAL CONSIDERATION

In light of recent developments in derivatives markets and as part of CFTC's reauthorization process, Congress should consider further exploring whether the current regulatory structure for energy derivatives, in particular for those traded in exempt commercial markets, provides adequately for fair trading and accurate pricing of energy commodities.

RECOMMENDATIONS FOR EXECUTIVE ACTION

To improve the oversight and available information on energy futures trading, we recommend that the Acting CFTC Chairman take the following three actions:

- reexamine the classifications in the COT reports to determine if the commercial and noncommercial trading categories should be refined to improve the accuracy and relevance of public information provided to the energy futures markets;
- explore ways to routinely maintain written records of inquiries into possible improper trading activity and the results of these inquiries to more fully determine the usefulness and extent of CFTC's surveillance, antifraud, and antimanipulation authorities; and
- examine ways to more fully demonstrate the effectiveness of CFTC enforcement activities by developing additional outcome-related performance measures that more fully reflect progress in meeting the program's overall goals.

APPENDIX I: SCOPE AND METHODOLOGY

To examine trends and patterns of trading activity in the energy derivatives markets and physical markets, we analyzed data on futures, spot, and over-the-counter (OTC) derivative markets. We gathered information on spot prices for crude oil, unleaded gasoline, heating oil, and natural gas from the U.S. Department of Energy's Energy Information Administration (EIA). We obtained daily futures settlement prices and average daily volume data for the four commodities from the New York Mercantile Exchange, Inc. (NYMEX). We collected data on the size of the global OTC commodity derivatives market—including energy, but excluding precious metals—from the Bank for International Settlements. We also obtained information on the numbers of participants and outstanding positions in energy futures markets by different categories of traders from the Commodity Futures Trading Commission (CFTC). These CFTC data cover the period from July 2003 through December 2006. We determined that data from these sources were sufficiently reliable for the purposes of this report.

We used monthly averages of the EIA spot prices and NYMEX futures prices to depict price trends over the past 20 years and illustrate the strong relationship between spot and futures prices. Also, we adjusted the prices to remove the effects of inflation so that prices would be comparable across years. We also adjusted the prices using monthly deflation factors that we derived from the seasonally adjusted implicit price deflator for gross domestic product from the Bureau of Economic Analysis, as of February 28, 2007.

We used the futures price data obtained from NYMEX to calculate the volatility of energy futures prices. These data covered the period from January 1987 through December 2006 for crude oil, unleaded gasoline, and heating oil. The period for natural gas was from April 1990 through December 2006, when that contract began trading on NYMEX. We calculated the historical volatility of the futures prices as the standard deviation of the natural logarithm of relative changes in daily settlement prices. Monthly volatility figures were calculated from the trading days of each month and expressed on an annual basis. We annualized the monthly figures by multiplying daily volatility by the square root of 250, which represents an approximation of the number of trading days in a year. We also calculated annual volatility for each of the four commodities as the average of the monthly mean volatilities. We used the front month futures contract—that is, the nearest traded contract month—because it is the most frequently used maturity for measuring price and volatility and is the most heavily traded contract.

To identify the opinions of market participants and analysts about the effect of energy derivatives trading on prices, we interviewed officials from CFTC, the Federal Energy Regulatory Commission, and EIA; managers from trading facilities, including NYMEX and the IntercontinentalExchange (ICE); academics knowledgeable about energy and finance; and market participants representing investment banks, hedge funds, and oil producers and refiners. We selected banks to interview on the basis of their perceived level of involvement in energy markets. The hedge funds we interviewed were identified through the assistance of the Managed Funds Association—a membership organization representing the hedge fund industry—which contacted its members involved in energy trading to identify hedge funds who were willing to be interviewed. We selected oil producers and refiners on the basis of their size and role in U.S. energy markets. We also gathered information from several trade associations, including those representing users of energy commodities, and interviewed former CFTC officials. Although we gathered the views of a wide range of market participants and observers, these sources do not necessarily represent the views of all market participants and observers. We also reviewed studies by governmental and nongovernmental observers, including CFTC; NYMEX; the Senate's Permanent Subcommittee on Investigations, Committee on Homeland Security and Governmental Affairs; and a report prepared for the attorneys general of four midwestern states. In addition, we reviewed public statements from relevant government officials, such as the current Federal Reserve chairman and his predecessor. To understand the effects of supply and demand conditions in the physical energy markets, we examined data and analysis from EIA and prior GAO reports.

To examine CFTC's resources and authority for protecting market users from fraudulent, manipulative, and abusive practices in the trading of energy futures contracts, we describe CFTC's current and past regulatory authority and approach by reviewing the Commodity Exchange Act (CEA), as amended; CFTC's President's Budget and Performance Plan for fiscal year 2007; CFTC's 2004

Appendix I: Scope and Methodology

Annual Report; and other information from CFTC. We obtained information on CFTC's regulatory role and the exempt commercial and OTC markets from officials at CFTC, EIA, and NYMEX and from market participants. In addition, we reviewed information on CFTC's regulatory role contained in the *Federal Register* and congressional hearing testimony. To describe the concerns regarding OTC derivative trading and the scope of CFTC's regulatory authority, we obtained information from federal agency officials and an industry trade association. To describe the hedging and speculative trading of market participants, we reviewed various reports that addressed those concerns, and we interviewed several market participants.

To examine how CFTC monitors and detects market abuses in the trading of energy futures, and enforcement actions taken in response to identified abuses, we gathered information from officials at CFTC headquarters and the New York Regional Office. We reviewed CFTC regulations and other documents on its surveillance and enforcement programs and observed a CFTC monthly surveillance meeting. We gathered information from market participants and experts regarding CFTC's oversight activities. To examine CFTC's enforcement program and how CFTC coordinates with other regulators and authorities, we gathered and analyzed data on CFTC's enforcement cases, interviewed CFTC and other federal agency officials and staff on coordination activities and agreements, and reviewed CFTC Office of the Inspector General reports. We also reviewed the Office of Management and Budget's PART assessment of CFTC's enforcement program.

APPENDIX II: TYPES OF CONTRACTS AND TRANSACTIONS FOR ENERGY COMMODITIES IN THE PHYSICAL AND FINANCIAL MARKETS

Markets	Type of contract or transaction	Features
Physical	Spot	Bilateral over-the-counter (OTC) transactions for immediate delivery or near-term delivery and payment representing a specific price and location. Industry analysts publish price data gathered from market participants.
	Forward	Bilateral OTC transactions in which the seller agrees to deliver to the buyer a specified quantity and quality of an asset or a commodity at a specified date at an agreed-upon price or pricing formula and where delivery is contemplated.
Financial	Derivatives traded on U.S. regulated exchanges	Futures contracts are standardized contracts for a specific product at a specific location, where delivery is not usually made and contracts are offset prior to expiration Transactions are executed on an exchange regulated by the Commodity Futures Trading Commission (CFTC), such as the New York Mercantile Exchange. The exchange publicly disseminates price and other data. Options on futures contracts are contracts that give a buyer the right, but not the obligation, to buy or sell a specific quantity of futures contracts within a designated period at a designated price.
	Derivatives traded on foreign boards of trade subject to foreign regulation	As in the United States, standardized contracts for a specific product at a specific location, where delivery is not usually made and contracts are usually offset prior to expiration. Sales are executed on an exchange, such as the IntercontinentalExchange (ICE) Futures in London that is subject to regulation by the U.K. Financial Services Authority. Foreign boards of trade are able to provide direct access to U.S. market participants by obtaining "no-action" relief from CFTC staff.

Markets	Type of contract or transaction	Features
	Derivatives traded on exempt commercial markets	Standardized contracts for a specific product at a specific location, where delivery is not usually contemplated because contracts are usually offset prior to expiration or are cash settled and are based on prices from a regulated futures exchange or another source. Transactions are executed on an electronic trading platform, such as ICE, involving "eligible commercial entities." Exempt commercial markets may offer a clearing service for certain derivatives contracts. Exempt commercial markets may offer trading both in contracts that are subject to the Commodity Exchange Act and contracts that are not.
	Bilateral OTC derivatives	Derivatives contracts that are privately negotiated, bilateral contracts between eligible counter parties, often involving a swap dealer. The contracts are financially settled and are based on prices from a regulated futures exchange or another source. OTC swaps are a promise between two parties to make a series of payments to each other, of which at least one series is based on a commodity price. OTC options: OTC markets also offer options to buy or sell other assets.

Sources: CFTC and GAO.

Note: Forward contracts have characteristics that make them similar to futures derivative contracts traded in a financial market. Both contracts represent agreements in which one party agrees to purchase a specified amount of an economic good at a specified price from another party at a future date or during some future period. For the purposes of this table, we chose to place forward contracts in the physical market category, rather than the financial market category, because parties entering into forward contracts are more likely to have the intention of exchanging the commodity than are parties entering into futures contracts. Parties entering into derivatives contracts rarely carry out an exchange of the physical commodity, as their purpose in entering the transaction is to assume or offset price risk.

APPENDIX III: NEW YORK MERCANTILE EXCHANGE SURVEILLANCE AND ENFORCEMENT ACTIVITIES

NYMEX CONDUCTS SURVEILLANCE OF BOTH MARKET AND TRADING ACTIVITIES

Under CFTC regulations, NYMEX is responsible for establishing and enforcing rules governing its member conduct and trading, preventing market manipulation, ensuring that futures industry professionals meet qualifications, and examining members for financial strength. In carrying out these responsibilities, NYMEX officials told us that NYMEX's surveillance program is designed to monitor market and trade practices. They said that NYMEX relies on automated detailed information for each transaction to identify the buyer, seller, and clearing members who maintain customer accounts, and to identify whether a person is a member of NYMEX. The market surveillance activities focus on monitoring for possible manipulation by market participants. Specifically, NYMEX officials monitor large trader data, the "street book" speculative position limits and accountability levels, exemptions to speculative position limits, position concentrations, and the relationship between cash and futures prices.[67] A speculative position limit is the maximum net position that a market participant may hold in a specified contract month of a listed NYMEX contract, and is set by NYMEX. Market participants that are bona fide hedgers are eligible to apply for, and receive under certain conditions, limited exemptions from speculative position limits.

NYMEX officials said that they monitor speculative position limits, and if a limit is about to be reached or has been hit, they record the overage and contact the customer to find out if there is a logical explanation for the overage that is linked to a bona fide commercial exposure. They added that if there is a logical explanation, the customer may be allowed to keep the position and file for a formal exemption; however, if there is no logical explanation, then NYMEX officials will direct the position to be reduced. The officials said they follow such directives with a warning letter to the customer. If the position is not reduced, the officials said that another warning letter is then to be issued. As a final step, NYMEX could hold a hearing before its business conduct committee to deny the customer access to the exchange. However, this has never happened, according to NYMEX officials. They also said that customers may obtain exemptions to speculative position limits on a case-by-case basis and for a period of 12 months. The officials said that exemptions are not given if they could disrupt the markets, and that the exemptions are monitored for possible changing circumstances, such as reorganizations or Moody's downgrading a party's equity rating.

NYMEX also monitors trading by customers using multiple brokers. For example, NYMEX officials stated that large oil companies may use several brokers to trade—that is, each broker may trade using its own account to hide what it is doing from other companies. However, NYMEX officials said that all of the company's accounts are aggregated, and that its staff will analyze the trading and large trader data and contact the customer if there are any surveillance issues. In addition, NYMEX has established the maximum daily trading limits for each commodity contract. These limits are the maximum price advance or decline from the prior day's settlement price and, if exceeded, trading stops for a period. NYMEX officials said that the exchange has changed these limits over the years, and the officials could not recall the last time a trading limit was reached. Furthermore, the limits were completely eliminated for NYMEX's New York Commodity Exchange (COMEX) division, which trades precious metals. According to NYMEX officials, suspending trading would give traders time to count and balance their positions before resuming trading. Unfortunately, the officials continued, if trading on NYMEX is suspended, the price discovery mechanism on which the OTC and cash markets depend is also suspended. They added that CFTC does not provide any requirements for price limits and, in fact, favors ongoing price discovery.

In addition to market surveillance, NYMEX conducts trade practice surveillance. According to NYMEX officials, this surveillance focuses on persons who handle contract orders either on the trading floor or electronically, as well as on persons and firms engaging in proprietary trading for their own accounts. NYMEX seeks to identify trading practice abuses, such as prearranged trading, front running, providing tips on proprietary information, and accommodating trades. Prearranged trading is the noncompetitive trading between brokers in accordance with an expressed or implied agreement or understanding and is a violation of CEA and CFTC regulations. Front running is taking a futures or option position based on, for example, a customer order in the same or related future or option. This practice is also known as trading ahead. Accommodation trading is noncompetitive trading entered into by a trader, usually to assist another with illegal trades. NYMEX officials stated that trade practice surveillance information may be used as part of market surveillance, but this surveillance is not as focused on price movements and involves different types of monitoring, such as physically observing floor trading by people entering orders, and, in effect, is similar to "police on the beat." The officials added that with their recent use of the Chicago Mercantile Exchange's Globex electronic trading system to trade energy futures, their trade practice surveillance system is changing, with more emphasis on monitoring access to, and activity on, the Globex system in NYMEX contracts.

NYMEX USES INFORMATION FROM SURVEILLANCE ACTIVITIES AND OTHER SOURCES FOR ENFORCEMENT CASES

As a self-regulatory organization (SRO), NYMEX has the authority to pursue instances of suspected manipulation or other attempts of fraudulent or abusive trading. NYMEX officials stated that the exchange conducts its own enforcement activities, and CFTC expects NYMEX as an SRO to handle issues relating to exchange members. However, NYMEX will request CFTC's assistance if needed, especially for issues relating to nonexchange members. If NYMEX officials become aware of potentially abusive practices outside of their authority, they will notify the appropriate federal regulator, such as CFTC. Information from NYMEX surveillance activities and other sources is used to investigate potential abuses. Other sources of information include referrals from CFTC, traders, and customers. NYMEX officials investigate these referrals; if there is evidence of wrongdoing, they may open an investigation case. For each case that is pursued, they record information, including the source of the referral and investigation activities. Interviews conducted during an investigation may be taped and transcribed. From January 2000 through May 2006, NYMEX opened 706 investigations. However, if a referral did not result in a case that was pursued, NYMEX does not document how each referral was handled. For example, if a referral regarding a trade resulted in a NYMEX official making a telephone call and finding that there was no apparent violation, NYMEX officials said staff would not create a written record to log how the referral was handled or the result of the inquiry.

NYMEX officials told us that when NYMEX pursues a case, such as prearranged trading, the case is brought before the exchange's business conduct committee (BCC).[68] According to NYMEX, the BCC (which includes three public members and other committee members), is structured in a manner analogous to a grand jury proceeding and determines, on the basis of the evidence in an investigative report and any written response from the person accused in the case, whether there is a reasonable basis to believe an exchange rule violation occurred. NYMEX officials told us that a BCC meeting is scheduled for every other month; however, in some months, there are no cases to discuss and no meetings are held. They added that the BCC hears about two or three dozen cases annually. Once a case is heard, the BCC may then direct the compliance department to issue a complaint, or, in the case of minor violations, a written warning. The person named in the complaint has 10 days to respond, and he or she may make a settlement offer at any time prior to the conclusion of a hearing. Settlement offers must be approved by NYMEX's board of directors. From January 2000 through May 2006, NYMEX opened more than 700 investigations, most involving trading violations, and, of those, 125 were BCC-issued complaints (see table 2).[69]

Settlement of NYMEX complaints may result in settlement offers, fines, or disciplinary actions. Settlement offers exceeding $25,000, or cases contested by the respondent, are referred to NYMEX's adjudication committee for settlement consideration or for a full disciplinary hearing. The adjudication committee is authorized to conduct hearings where the facts of the case are presented and argued by the respondent or their attorney and exchange compliance counsel. At the conclusion of the hearing, the committee issues a decision regarding any sanctions.

Table 2. Number of NYMEX Enforcement Cases Opened, Complaints Issued, Settlement, and Hearings, January 2000–May 2006

Year	Number of investigations opened[a]	Number of complaints issued[b]	Number of respondents[c]	Number settled at BCC[d]	Number settled at Adjudication[e]	Number contested at hearings and adjudication[f]	Hearing results[g]
2000	101	19	21	4	13	2	Dismissed. Charges affirmed on appeal; fine and suspension reduced.[b]
2001	88	21	28	3	22	0	
2002	115	21	27	9	16	0	
2003	123	16	22	9	14	1	Rule violations found; fine.
2004	111	18	28	11	11	0	
2005	97	24	37	3	23	1	Decision pending.
2006	71	6	9	2	4	0	

Source: GAO analysis of NYMEX data.

Note: The number of complaints and respondents do not correspond to the settled and hearing numbers on a one-for-one basis.

[a] Number of investigations opened is based on investigations opened within the calendar years 2000, 2001, 2002, 2003, 2004, 2005 and through May 2006 (and taken from an annual review of the Compliance Department conducted by NYMEX's legal staff for a report to the Board). Not all inquiries become formal investigations. For instance, in 2005, the market surveillance area reported 887 "cases" on their rule violation and e-inquiry log. The vast majority of these were routine position limit reviews, exchange for physical and exchange of futures for swaps inquiries, and unreported reviews that never became formal investigations. The trade practice area logs "inquires," which may or may not evolve into formal investigations.

[b] Number of complaints issued represents the number of complaints issued by the BCC during the year specified, but only for investigations opened in one of the target years (2000 through May 2006).

[c] Number of respondents represents the number of individuals charged in the various BCC cases.

[d] Number settled at BCC represents the number of individuals who settled cases from a target year before the BCC.

[e] Number settled at adjudication represents the number of individuals who settled cases from a target year before the adjudication committee, which has authority similar to a judge and jury.

[f] Number contested at hearings and adjudication indicates any cases that were contested (rather than settled) and proceeded to a hearing or adjudication.

[g] Hearing results also encompass the results of adjudications.

Sanctions can include a cease-and-desist order, a fine of up to $250,000 for each rule violation, suspension, or expulsion from membership. For example, NYMEX settled a $100,000 case on Morgan Stanley in 2002, a $2.5 million case on BP Product North America, Inc., in 2003, and a $300,000 case on Shell in 2005. NYMEX officials stated that no one has been expelled from the exchange since 1998 for failure to pay a fine. According to the officials, the adjudication committee is scheduled to meet at least once a month. They added that about 40 percent of the cases are resolved at adjudication and cases rarely go to a full hearing. If a hearing does occur, the decisions can be appealed to NYMEX's appeals committee, which makes a final determination within the exchange. Cases then can be appealed further to CFTC, but cases involving an appeal of an exchange appeals committee decision rarely are contested or proceed to a hearing. NYMEX officials said that the commission rarely, if ever, overturns a NYMEX ruling. NYMEX publishes its final disciplinary actions of the exchange. All settlements or adjudications are published in its monthly publication *The Open Interest* (formerly, *Barrels, Bars and BTUs*) and sent to the National Futures Association, where they are included in the publicly accessible disciplinary log called "BASIC," which contains reports from all U.S. futures exchanges and from CFTC. Warning letters are not reported, but they are used internally in prosecuting disciplinary cases. NYMEX also reports its enforcement actions to CFTC.

To maintain its designation as a contract market, NYMEX must demonstrate to CFTC its capacity to comply with the CEA's core principles. In addition, CFTC conducts rule enforcement reviews and publicly reports on how NYMEX exercises its enforcement authority and other areas of operations. In 2004, CFTC reported that NYMEX's disciplinary program provided reasonable sanctions for a majority of the cases where the exchange took disciplinary action, and that its dispute resolution program had fair and equitable procedures. The report was generally positive and reported that NYMEX's procedures provided for the recording and safe storage of trade information. Furthermore, NYMEX's surveillance of trade practices was deemed to be adequate, with thorough and well-documented investigations. NYMEX officials told us that a CFTC rule enforcement review was recently initiated at the exchange.

APPENDIX IV: COMMODITY FUTURES TRADING COMMISSION'S ENERGY-RELATED ENFORCEMENT ACTIONS, AUGUST 2001 - SEPTEMBER 2006

Table 3 reflects information obtained from CFTC showing the energy-related enforcement actions it took by CFTC from August 2001 through September 2006. The enforcement cases were against individuals and companies, and the information used to initiate the investigation originated from both within CFTC and from outside of the commission. The actions mostly focused on attempts to manipulate energy commodity prices through alleged attempts of false reporting; some also involved alleged wash sales or trades—transactions intended to give the appearance that purchases and sales have been made, without incurring market risk or changing the trader's market position, prearranged trading, and recordkeeping violations. CFTC's information also shows a wide range of civil monetary penalties.

Table 3. Energy-Related Enforcement Actions Filed by CFTC, August 2001–September 2006

Date case filed	CFTC enforcement case	Initial source of information CFTC used to initiate the investigation (CFTC/external/individual)	Reason for charges	Civil monetary penalties
09/2006	Dominion Resources, Inc.	CFTC	False reporting	$4.3 million
06/2006	BP Products North America, Inc.	External	Manipulation, cornering the market, and attempted manipulation	In litigation
01/2006	Shell Trading US Company and Shell International Trading and Shipping Co., Nigel Catterall	External	Prearranged trading	$300,000
09/2005	Joseph Foley	CFTC/External	Attempted manipulation and false reporting	$350,000
05/2005	Brion Scott McKenna	CFTC	Manipulation	Registration revoked
04/2005	Andrew Richmond	CFTC/External	Attempted manipulation and false reporting	$60,000
02/2005	Christopher McDonald, Michael Whalen, and Paul Atha	CFTC/External	Attempted manipulation and false reporting	$350,000 $200,000 In litigation
02/2005	Matthew Reed, Darrell Danyluk, Shawn Mclaughlin, and Concord Energy	CFTC	Attempted manipulation and false reporting	In litigation $350,000 $450,000 $800,000
02/2005	Jeffrey A. Bradley, Robert Martin	External	Attempted manipulation and false reporting	In litigation In litigation
02/2005	Denette Johnson, Courtney Cubbison Moore, John Tracy, Robert Harp, Anthony Dizona, and Kelly Dyer	External	Attempted manipulation and false reporting	All in litigation
02/2005	Michael Whitney	External	Attempted manipulation and false reporting	In litigation

Table 3. Continued

Date case filed	CFTC enforcement case	Initial source of information CFTC used to initiate the investigation (CFTC/external/individual)	Reason for charges	Civil monetary penalties
12/2004	Mirant Americas Energy Marketing, L.P.	CFTC/External	Attempted manipulation and false reporting	$12.5 million
11/2004	Cinergy Marketing and Trading, L.P.	CFTC/External	False reporting	$3 million
11/2004	BP Energy Company	CFTC	Wash trades	$100,000
08/2004	Byron Biggs	CFTC	Wash trades	$30,000
07/2004	United Energy and Dana Christopher Bray	CFTC/External	Recordkeeping violations	$33,000
07/2004	NRG Energy, Inc.	CFTC	False reporting	$2 million
07/2004	Coral Energy Resources, L.P.	External	Attempted manipulation and false reporting	$30 million
07/2004	Western Gas Resources	CFTC/External	Attempted manipulation and false reporting	$7 million
01/2004	Robert Benjamin Harmon, Jr.	External	Wash trades	$7,500
05/2004	Joseph Knauth	CFTC	Wash trades	$25,000
05/2004	Enron Corporation and Hunter S. Shively	CFTC/External	Manipulation or attempted manipulation. Enron only: Operating illegal futures exchange and trading an off-exchange agricultural futures contract	$35 million $300,000

Table 3. Continued

Date case filed	CFTC enforcement case	Initial source of information CFTC used to initiate the investigation (CFTC/external/individual)	Reason for charges	Civil monetary penalties
01/2004	Calpine Corporation	CFTC	False reporting	$1.5 million
01/2004	ONEOK, Inc., ONEOK Energy Marketing and Trading Co., L.P.	Individual	False reporting	$3 million
01/2004	Entergy Koch Trading L.P.	External	False reporting	$3 million
01/2004	e prime	CFTC/External	Attempted manipulation and false reporting	$16 million
01/2004	Aquila Merchant Services	CFTC/External	Attempted manipulation and false reporting	$26.5 million
11/2003	CMS Marketing Services and Trading; CMS Field Services	External	Attempted manipulation and false reporting	$16 million
11/2003	Reliant Energy Services, Inc.	External	Attempted manipulation and false reporting, and wash sales	$18 million
09/2003	William Taylor	CFTC	Attempted Manipulation	$155,000
09/2003	Michael Garber, NYMEX floor broker	External	Wash sales, reported non-bona fide prices, and noncompetitive trading	$7,500
09/2003	American Electric Power Company (AEP) and AEP Energy Services (AEPES)	CFTC/External	Attempted manipulation and false reporting	$30 million
09/2003	Duke Energy Trading and Marketing, L.L.C.	External	Attempted manipulation and false reporting	$28 million
07/2003	Enserco Energy	CFTC	Attempted manipulation and false reporting	$3 million
07/2003	Williams Companies and Williams Energy Marketing and Trading	CFTC/External	Attempted manipulation and false reporting	$20 million

Table 3. Continued

Date case filed	CFTC enforcement case	Initial source of information CFTC used to initiate the investigation (CFTC/external/individual)	Reason for charges	Civil monetary penalties
07/2003	W. D. Energy Services, Inc. (Encana)	External	Attempted manipulation and false reporting	$20 million
03/2003	Christopher Chapman	External	Fraudulent trading	$240,000
03/2003	El Paso Merchant Energy	CFTC/External	Attempted manipulation and false reporting	$20 million
12/2002	Dynegy Marketing and Trade and West Coast Power	CFTC	Attempted manipulation and false reporting	$5 million
08/2001	Robert Kristufek	CFTC	Attempted manipulation	$155,000
08/2001	Thomas Johns, Michael Griswold, and Avista Energy	CFTC	Attempted manipulation	$50,000 $110,000 $2.1 million

Source: CFTC.

Note: According to CFTC, AEPES entered into a deferred prosecution agreement with the Department of Justice and the U.S. Attorney's Office for the Southern District of Ohio to avoid federal criminal charges. The agreement requires AEPES to pay a $30 million criminal penalty to resolve an investigation into AEPES' false reporting of natural gas trades. In addition, AEP accepted a settlement agreement with the Federal Energy Regulatory Commission (FERC) to resolve an investigation into the natural gas storage and transportation activities of two intrastate pipeline units formerly owned by AEP and AEP-affiliated marketers. The FERC settlement requires AEP to pay a $21 million civil penalty and adopt a compliance plan to prevent future violations. The total settlement with the U.S. government was $81 million.

ADDITIONAL READING

Energy Markets: Factors Contributing to Higher Gasoline Prices. GAO-06-412T. Washington, D.C.: February 1, 2006.

Natural Gas and Electricity Markets: Federal Government Actions to Improve Private Price Indices and Stakeholder Reaction. GAO-06-275. Washington, D.C.: December 15, 2005.

SEC and CFTC Penalties: Continued Progress Made in Collection Efforts, but Greater SEC Management Attention Is Needed. GAO-05-670. Washington, D.C.: August 31, 2005.

Mutual Fund Industry: SEC's Revised Examination Approach Offers Potential Benefits, but Significant Oversight Challenges Remain. GAO-05-415. Washington, D.C.: August 17, 2005.

National Energy Policy: Inventory of Major Federal Energy Programs and Status of Policy Recommendations. GAO-05-379. Washington, D.C.: June 10, 2005.

Motor Fuels: Understanding the Factors That Influence the Retail Price of Gasoline. GAO-05-525SP. Washington, D.C.: May 2005.

Mutual Fund Trading Abuses: SEC Consistently Applied Procedures in Setting Penalties, but Could Strengthen Certain Internal Controls. GAO-05-385. Washington, D.C.: May 16, 2005.

Financial Regulation: Industry Changes Prompt Need to Reconsider U.S. Regulatory Structure. GAO-05-61. Washington, D.C.: October 6, 2004.

Natural Gas: Domestic Nitrogen Fertilizer Production Depends on Natural Gas Availability and Prices. GAO-03-1148. Washington, D.C.: September 30, 2003.

SEC and CFTC Fines Follow-up: Collection Programs Are Improving, but Further Steps Are Warranted. GAO-03-795. Washington, D.C.: July 15, 2003.

Natural Gas: Analysis of Changes in Market Price. GAO-03-46. Washington, D.C.: December 18, 2002.

SEC and CFTC: Most Fines Collected, but Improvements Needed in the Use of Treasury's Collection Service. GAO-01-900. Washington, D.C.: July 16, 2001.

Energy Markets: Results of Studies Assessing High Electricity Prices in California. GAO-01-857. Washington, D.C.: June 29, 2001.

Commodity Exchange Act: Issues Related to the Regulation of Electronic Trading Systems. GAO/GGD-00-99. Washington, D.C.: May 5, 2000.

CFTC and SEC: Issues Related to the Shad-Johnson Jurisdictional Accord. GAO/GGD-00-89. Washington, D.C.: April 6, 2000.

Financial Regulatory Coordination: The Role and Functioning of the President's Working Group. GAO/GGD-00-46. Washington, D.C.: January 21, 2000.

The Commodity Exchange Act: Issues Related to the Commodity Futures Trading Commission's Reauthorization. GAO/GGD-99-74. Washington, D.C.: May 5, 1999.

CFTC Enforcement: Actions Taken to Strengthen the Division of Enforcement. GAO/GGD-98-193. Washington, D.C.: August 28, 1998.

OTC Derivatives: Additional Oversight Could Reduce Costly Sales Practice Disputes. GAO/GGD-98-5. Washington, D.C.: October 2, 1997.

CFTC/SEC Enforcement Programs: Status and Potential Impact of a Merger. GAO/T-GGD-96-36. Washington, D.C.: October 25, 1995.

Financial Market Regulation: Benefits and Risks of Merging SEC and CFTC. GAO/T-GGD-95-153. Washington, D.C.: May 3, 1995.

Energy Security and Policy: Analysis of the Pricing of Crude Oil and Petroleum Products. GAO/RCED-93-17. Washington, D.C.: March 19, 1993.

Securities and Futures: How the Markets Developed and How They Are Regulated. GAO/GGD-86-26. Washington, D.C.: May 15, 1986.

REFERENCES

[1] Our analysis of energy prices and energy financial markets generally is limited to the period from January 2002 through December 2006. A "derivative" is a financial instrument, traded on- or off-exchange, the price of which for energy directly depends on the value of one or more underlying energy commodities. Derivatives involve the trading of rights or obligations on the basis of the underlying product, but they do not directly transfer property.

[2] CFTC's technical definition of a "futures contract" encompasses the following characteristics: (1) the contract price is determined at initiation of the contract; (2) the contract obligates each party to the contract to fulfill the contract at the specified price; (3) the contract is used to assume or shift price risk; and (4) the delivery obligation may be satisfied by delivery or offset (i.e., liquidating a purchase of futures contracts through the sale of an equal number of contracts of the same delivery month, or liquidating a short sale of futures through the purchase of an equal number of contracts of the same delivery month). A futures contract is a type of derivative.

[3] CFTC defines "hedge fund" as a private investment fund or pool that trades and invests in various assets, such as securities, commodities, currency, and derivatives, on behalf of its clients, typically wealthy individuals. In a "commodity index fund," prices are tied to the price of a basket of various commodity futures.

[4] See section 3 of the Commodity Exchange Act, 7 U.S.C. § 5 (2004).

[5] The LTRS includes the daily reports filed with CFTC showing the futures and options positions of traders that hold positions at or above specific exchange or CFTC-set reporting levels. Commodity traders or brokers that carry these accounts must make daily reports about the size of the position

by commodity, by delivery month, and by whether the position is controlled by a commercial or noncommercial trader. Commercial participants generally are those that are engaged in business activities—including producing, merchandising, or processing a cash commodity or managing risk—that hedge using the futures or options markets. Noncommercial participants do not have an interest in the underlying commodity but trade in the energy futures markets to realize a profit.

[6] The CEA defines "exempt commodity" as a commodity that is "not an excluded commodity or an agricultural commodity." 7 U.S.C. § 1a(14). In practice, this definition primarily encompasses energy and metal commodities.

[7] In October 2005, NYMEX began to offer a new futures contract for reformulated gasoline blendstock known as "RB." This new contract traded alongside the existing gasoline contract known as "HU" until January 2007, when NYMEX discontinued trading in that contract.

[8] CFTC economists used the term "managed money traders" to describe a large category of speculative traders. See Michael S. Haigh, Jana Hranaiova, and James A. Overdahl, "Price Dynamics, Price Discovery and Large Futures Trader Interactions in the Energy Complex," U.S. Commodity Futures Trading Commission, Office of the Chief Economist, April 28, 2005, draft working paper available on CFTC's Web site. As stated in the paper, the views expressed therein are those of the authors and do not reflect the views of CFTC or its staff.

[9] CPOs are individuals or firms in businesses similar to investment trusts or syndicates that solicit or accept funds, securities, or property for the purpose of trading futures or commodity options. CTAs are individuals or firms that, for pay, issue analyses or reports concerning commodities, including the advisability of trading futures or commodity options. For the purposes of the working paper, the authors used the term "managed money traders" to include all registered CPOs and CTAs.

[10] To account for the effects of inflation on prices, prices are adjusted to reflect prices in the base year of 2006.

[11] This scenario assumes that minimal costs are associated with holding the oil over the 2-week period.

[12] Futures Trading Practices Act of 1992 (Pub. L. No. 102-546 (Oct. 28, 1992)). Among other things, this act added a new provision to the CEA authorizing the commission, by rule, regulation, or order, to exempt any agreement, contract, or transaction, or class thereof, when entered into between "appropriate persons" from the exchange-trading, or any other,

requirement of the act (other than the provision establishing CFTC's jurisdiction). Pub. L. No. 102-546 § 502. In April 1993, CFTC promulgated a final order generally exempting from the CEA qualifying energy contracts entered by commercial participants and certain other specified entities. 58 Fed. Reg. 21286 (Apr. 20, 1993).

[13] Pub. L. No. 106-554 § 1(a)(5), title 1 §§ 103-106 (Dec. 21, 2000).

[14] Another type of futures exchange that is subject to CFTC regulatory oversight is a derivatives transaction execution facility, which is a trading facility that limits access to mostly institutional traders rather than retail traders. Like a futures market, a derivatives transaction execution facility must register with CFTC but is subject to less regulation. To date, no exchange has applied to register as such a facility.

[15] President's Working Group on Financial Markets, *Over-the-Counter Derivatives Markets and the Commodity Exchange Act* (Nov. 9, 1999). Members of the President's Working Group on Financial Markets include the Chairman of CFTC, the Secretary of the Treasury, the Chairman of the Board of Governors of the Federal Reserve, and the Chairman of SEC.

[16] GAO, *Financial Regulation: Industry Changes Prompt Need to Reconsider U.S. Regulatory Structure,* GAO-05-61 (Washington, D.C.: Oct. 6, 2004).

[17] CFTC's other major operating units are the Office of the Chief Economist, Office of the General Counsel, and Office of the Executive Director. A derivatives clearing organization is a clearinghouse or similar organization that enables each party to a transaction to substitute the credit of the clearinghouse for the credit of the parties, provides for the settlement or netting of obligations from the transaction, or otherwise provides services mutualizing or transferring the credit risk from the transaction.

[18] Futures commission merchants are individuals, associations, partnerships, corporations, and trusts that solicit or accept orders for the purchase or sale of any commodity for future delivery on or subject to the rules of any exchange and that accept payment from or extend credit to those whose orders are accepted.

[19] As of January 1, 2007, CFTC closed its Minneapolis field office.

[20] Because energy commodities are related, specific types of events or conditions may affect all four energy commodities in a similar fashion. For example, the supply and demand fundamentals for crude oil have a direct effect on the supply, demand, and price of gasoline and heating oil because they are refined from crude oil. Crude oil market fundamentals also may affect natural gas because some consumers can use it as a substitute for crude oil products. If the price of crude oil rises, demand for natural gas as a

substitute may rise, thereby increasing its price. However, natural gas prices are not always closely related to crude oil prices.

[21] The OPEC members are Algeria, Angola, Indonesia, Iran, Iraq, Kuwait, Libya, Nigeria, Qatar, Saudi Arabia, the United Arab Emirates, and Venezuela. Http://www.eia.doe.gov/emeu/cabs/AOMC/Overview.html.

[22] GAO, *Motor Fuels: Understanding the Factors That Influence the Retail Price of Gasoline*, GAO-05-525SP (Washington, D.C.: May 2005).

[23] GAO, *Energy Markets: Factors Contributing to Higher Gasoline Prices*, GAO-06-412T (Washington, D.C.: Feb. 1, 2006); and GAO-05-525SP.

[24] GAO-06-412T.

[25] GAO, *Gasoline Markets: Special Gasoline Blends Reduce Emissions and Improve Air Quality, but Complicate Supply and Contribute to Higher Prices*, GAO-05-421 (Washington, D.C.: June 17, 2005).

[26] GAO-05-525SP.

[27] Mark N. Cooper, *The Role of Supply, Demand and Financial Commodity Markets in the Natural Gas Price Spiral*, prepared for Midwest Attorneys General Natural Gas Working Group (Illinois, Iowa, Missouri, and Wisconsin: March 2006).

[28] The CFTC report did not look at the effects of recent trends in volume or the number of large traders on prices.

[29] "Standard deviation" is a measure of the dispersion of a set of data around its mean. When used to measure volatility, standard deviation measures the dispersion of daily percentage price changes around the average percentage price change.

[30] CFTC collects data on traders holding positions at or above specific reporting levels set by the commission. This information is collected as part of CFTC's LTRS.

[31] Senate's Permanent Subcommittee on Investigations, Committee on Homeland Security and Governmental Affairs, *The Role of Market Speculation in Rising Oil and Gas Prices: A Need to Put the Cop Back on the Beat*, S. Prt. 109-65, 109th Cong., 2nd Sess. (June 27, 2006).

[32] The Bank for International Settlements is an international organization that fosters international monetary and financial cooperation and serves as a bank for central banks.

[33] The "notional amount" is the amount upon which payments between parties to certain types of derivatives contracts are based. The notional amount is not exchanged between the parties, but instead represents a hypothetical underlying quantity upon which payment obligations are computed. The

Bank for International Settlements data on OTC derivatives include forwards, swaps, and options.
[34] The CEA antimanipulation and antifraud prohibitions do not apply to excluded derivative and swap transactions. 7 U.S.C. § 2(d)–(g). The antimanipulation prohibitions apply to off-exchange transactions in exempt commodities. The antifraud provisions apply to transactions in exempt commodities only under particular circumstances.
[35] 17 C.F.R. § 36.3; see 7 U.S.C. § 2(h)(4)(D).
[36] On June 22, 2007, CFTC published for comment a proposed rule that, among other things, is intended to clarify that a person holding or controlling reportable positions on a futures exchange must keep records of and make available to CFTC information about all of the trader's transactions in the commodity reported, including any transactions in the exempt commercial markets, such as OTC energy derivatives. 72 Fed. Reg. 34413 (June 22, 2007).
[37] 7 U.S.C. § 7. Part 38 of the CFTC rules sets forth the procedures and criteria for designation as a contract market. Among other things, these procedures and criteria include guidance on compliance with CEA designation criteria and acceptable practices in compliance with the core principles. See 17 C.F.R. Part 38.
[38] 7 U.S.C. § 7a-2(c)(3).
[39] 17 C.F.R. Parts 155, 166.
[40] Under a special call for 2(h)(3) (exempt commodity) markets, ICE, which is not subject to the oversight CFTC has over futures markets such as NYMEX, is providing large trader data to CFTC. The COT report was first published monthly in 1962. Since 1995, it has been available for free at CFTC's Web site; since 2000, it has been published weekly.
[41] 71 Fed. Reg. 35627, 35630-31 (June 21, 2006).
[42] 17 C.F.R. § 36.3. CFTC's antifraud authority under the CEA applies only to transactions within the commission's authority. Therefore, CFTC's antifraud authority would not apply to cash or forward transactions on exempt commercial markets.
[43] As discussed in footnote 54, a ruling by one federal appellate court means that the CEA antifraud provision may not apply to off-exchange transactions conducted on a principal-to-principal basis.
[44] These requirements are contained in CFTC rule 36.3.
[45] 7 U.S.C. § 2(h)(4)(D).
[46] 7 U.S.C. § 2(h)(5).

[47] In a recent CFTC complaint filed against Amaranth Advisors, LLC; Amaranth Advisors (Calgary), ULC; and Brian Hunter, CFTC alleges that the defendants attempted to manipulate the price of natural gas contracts on NYMEX in 2006. *CFTC v. Amaranth Advisors, LLC*, '07 CIV 6682 (SD NY, July 25, 2007).
[48] ICE also trades financially settled contracts.
[49] 7 U.S.C. § 2(h)(3). The term "eligible commercial entity" is defined at 7 U.S.C. § 1a(11). In general, these participants are entities such as financial institutions, commodity pools, or large businesses that, by virtue of their regulatory or financial status, are permitted to engage in transactions not available to other participants, such as retail customers.
[50] 7 U.S.C. § 2(h)(2),(4). As we have previously noted in this report, if the electronic trading facility upon which these exempt contracts are traded becomes a significant price discovery market, it may be subject to CFTC rules on the timely dissemination of pricing data and trading volume and information. Also, a facility relying on the exemption must notify the commission of its intent to operate; provide the name of the facility; describe the types of commodity categories being traded; identify its clearing facility, if any; certify that the facility will comply with the terms of the exemptions; certify that the owners of the trading facility are not otherwise statutorily disqualified under the CEA; either provide the commission with real-time access to its trading system and protocols, or provide CFTC with such reports as it may request; maintain books and records for 5 years; agree to provide the commission with specific information on a special call basis; agree to submit to CFTC's subpoena authority; agree to comply with all applicable laws and require the same of its participants; and not represent that the facility is registered with or in any way recognized by CFTC. 17 C.F.R. § 36.3; see 7 U.S.C. § 2(h)(5).
[51] Eva Gutierrez, *A Framework for the Surveillance of Derivatives Activities*, International Monetary Fund Working Paper WP/05/61 (Washington, D.C.: March 2005).
[52] *Over-the-Counter Derivatives Markets and the Commodity Exchange Act.*
[53] As stated by CFTC, the purpose of the proposed regulation is to make it explicit that persons holding or controlling reportable positions on a reporting market must retain books and records and make available to the commission upon request any pertinent information with respect to all other positions and transactions in the commodity in which the trader has a reportable position, including positions held or controlled or transactions executed over-the-counter or pursuant to sections 2(d), 2(g) or 2(h)(1)–(2)

of the CEA or part 35 of the commission's regulations, on exempt commercial markets operating pursuant to sections 2(h)(3)–(5) of the CEA, on exempt boards of trade operating pursuant to Section 5d of the CEA, and on foreign boards of trade (hereinafter referred to collectively as nonreporting transactions); and to make the regulation clearer and more complete with respect to hedging activity. The purpose of the amendments is to clarify CFTC's regulatory reporting requirements for such traders. 72 Fed. Reg. 34413.

[54] Section 4b of the CEA is CFTC's main antifraud authority. In a November 2000 decision, the 7th Circuit Court of Appeals ruled that CFTC only could use section 4b in intermediated transactions—those involving a broker. *Commodity Trend Service, Inc. v. CFTC*, 233 F.3d 981, 991-992 (7th Cir. 2000). As amended by the CFMA, the CEA permits off-exchange futures and options transactions that are done on a principal-to-principal basis, such as energy transactions pursuant to CEA sections 2(h)(1) and 2(h)(3). According to CFTC, House and Senate CFTC reauthorization bills introduced during the 109th Congress (H.R. 4473 and S. 1566) would have amended section 4b to clarify that Congress intends for CFTC to enforce section 4b in connection with off-exchange principal-to-principal futures transactions, including exempt commodity transactions in energy under section 2(h) as well as all transactions conducted on derivatives transaction execution facilities.

[55] Commodity Futures Trading Commission, *Performance and Accountability Report: Fiscal Year 2005* (Washington, D.C.: November 2005).

[56] The LTRS data also can be used to identify violations of speculative position limits. The CEA authorizes CFTC to impose limits on the size of speculative positions in futures markets to protect futures markets from excessive speculation that can cause unreasonable or unwarranted price fluctuations. Exchanges establish speculative limits for energy products. Violations of exchange speculative limits may be charged by the commission as violations of section 4a of the CEA, if the exchange rules have been approved by the commission.

[57] *In re Indiana Farm Bureau Cooperative Association,* [1982-1984 Transfer Binder] *Comm. Fut. L. Rep. (CCH) P21,796 at 27,281,* n.2 (CFTC: Dec. 17, 1982) (explaining that "[i]n order to prove a successful manipulation, it is necessary to demonstrate that the accused intentionally caused an 'artificial price,' that is, a price which does not reflect the market or economic forces of supply and demand.") One of the purposes of the CEA is to prevent market manipulation. *Curran v. Merrill Lynch, Pierce, Fenner*

and Smith, Inc., 622 F.2d 216, 235 (6th Cir. 1980) (noting the "Congressional intent [in enacting the act] to protect the public from fraud and price manipulation"). Sections 6(c), 6(d), and 9(a)(2) of the CEA, 7 U.S.C. §§ 9, 13b, 13(a)(2), make it illegal for any person to manipulate or attempt to manipulate the market price of any commodity, in interstate commerce, or for future delivery on or subject to the rules of any registered entity (including any contract market), or to corner or attempt to corner any such commodity or knowingly deliver or cause to be delivered false, misleading, or knowingly inaccurate reports concerning crop or market information or conditions that affect or tend to affect the price of any commodity in interstate commerce. As a result of the commission carrying out the statutory mandate through enforcement actions and administrative decisions, a body of federal and commission case law has emerged to define manipulation under the act. Generally, the following factors are assessed in manipulation cases: (1) that the accused had the ability to influence market prices, (2) that the accused specifically intended to do so, (3) that artificial prices existed, and (4) that the accused caused an artificial price. *In re Cox*, [1986-1987 Transfer Binder] Comm. Fut. L. Rep. (CCH) 23,786 at 34,061 (CFTC: July 15, 1987).

[58] A position is a long or short interest in the market in the form of one or more contracts.

[59] GAO, *Long-Term Capital Management: Regulators Need to Focus Greater Attention on Systemic Risk*, GAO/GGD-00-3 (Washington, D.C.: Oct. 29, 1999).

[60] Section 9(a)(2) of the CEA prohibits "(a)ny person to manipulate or attempt to manipulate the price of any commodity in interstate commerce, or for future delivery on or subject to the rules of any registered entity, or to corner or attempt to corner any such commodity or knowingly to deliver or cause to be delivered for transmission through the mails or interstate commerce by telegraph, telephone, wireless, or other means of communication false or misleading or knowingly inaccurate reports concerning crop or market information or conditions that affect or tend to affect the price of any commodity interstate commerce...."

[61] Section 9 of the CEA, 7 U.S.C. § 13(a)(2), makes it a felony for any person to manipulate or attempt to manipulate the price of any commodity in interstate commerce as well as the prices of futures contracts.

[62] See Order to Show Cause and Notice of Proposed Penalties, FERC Docket No. IN07-26-000 (July 26, 2007).

[63] *CFTC v. BP Products North America, Inc.,* No. 06C 3503 (N.D. Ill. filed June 28, 2006).

[64] GAO reports cited in the PART assessment: *CFTC Enforcement: Actions Taken to Strengthen the Division of Enforcement,* GAO/GGD-98-193 (Washington, D.C.: Aug. 23, 1998); *SEC and CFTC Fines Follow-Up Collection Programs Are Improving, but Further Steps Are Warranted,* GAO-03-795 (Washington, D.C.: July 15, 2003); *Results Act: Observations on CFTC's Annual Performance Plan,* GAO/T-GGD-99-10 (Washington, D.C.: Oct. 8, 1998); and *Results Act: Observations on CFTC's Strategic Plan,* GAO/T-GGD-98-17 (Washington, D.C.: Oct. 22, 1997).

[65] GAO, *Results Oriented Government: GPRA Has Established a Solid Foundation for Achieving Greater Results,* GAO-04-594T (Washington, D.C.: Mar. 31, 2004).

[66] GAO, *Program Evaluation: OMB's PART Reviews Increased Agencies' Attention to Improving Evidence of Program Results,* GAO-07-67 (Washington, D.C.: Oct. 28, 2005).

[67] The street book is a daily record showing details of each futures and option transaction, including date, price quantity, market, commodity, future, and name of the person for whom the trade was made.

[68] There is a NYMEX BCC and COMEX BCC. A BCC meeting is scheduled to meet every month, alternatively for NYMEX and COMEX.

[69] NYMEX said that not all inquiries become formal investigations. For example, in 2005, the Market Surveillance area reported 887 cases, but the vase majority of these were routine position limit reviews, inquires about exchange for physical and exchange of futures for swaps, and unreported reviews that never became formal investigations. The Trade Practice area logs inquiries that may or may not evolve to formal investigations.

INDEX

A

academics, 60
access, 3, 10, 63, 66, 67, 81, 84
accountability, 44, 45, 52, 53, 65
accuracy, 6, 8, 33, 35, 57
acquisitions, 51
advocacy, 24
air quality, 23
Algeria, 82
alternatives, 19
amendments, 40, 85
Angola, 82
annual review, 69
appendix, 8
arbitrage, 14
argument, 37
Asia, 20
assessment, 32, 37, 48, 50, 51, 52, 53, 61, 87
assets, 10, 64, 79
auditing, 4
authority, vii, 1, 3, 6, 15, 31, 32, 35, 39, 40, 43, 49, 50, 51, 55, 60, 67, 69, 70, 83, 84, 85
availability, 37
averaging, 36

B

banking, 33, 39
banks, 4, 27, 35, 39, 60, 82
base year, 80
behavior, 51, 53
benefits, 52
blends, 23
Board of Governors, 39, 49, 81
BP, 51, 70, 72, 73, 87
BTUs, 70
butyl ether, 23
buyer, 63, 65

C

California, 78
Canada, 21
capacity, 22
case law, 86
causality, 55
central bank, 82
certification, 32
chemical, 23
Chicago, 4, 16, 38, 67
China, 20, 21
Chinese, 20
classification, 33
Clean Air Act, 23
clients, 79
collusion, 51
commerce, 86

commercial(s), 1, 6, 7, 10, 11, 12, 15, 27, 31, 33, 34, 35, 36, 37, 38, 39, 40, 41, 45, 55, 57, 61, 64, 66, 80, 81, 83, 84, 85
commercial bank, 39
Committee on Homeland Security, 27, 60, 82
commodity(ies), vii, ix, 1, 2, 3, 6, 7, 9, 10, 11, 12, 14, 15, 16, 20, 21, 23, 24, 27, 28, 30, 31, 32, 33, 34, 35, 36, 37, 38, 39, 43, 48, 50, 52, 56, 57, 59, 60, 63, 64, 66, 71, 79, 80, 81, 83, 84, 85, 86, 87
Commodity Exchange Act (CEA), ix, 1, 6, 15, 16, 31, 32, 33, 35, 36, 38, 39, 40, 44, 49, 50, 52, 60, 64, 67, 78, 79, 80, 81, 83, 84, 85, 86
commodity futures, 15, 32, 34, 35, 43, 52, 79
Commodity Futures Trading Commission (CFTC), vii, ix, 1, 2, 3, 6, 7, 8, 10, 12, 15, 16, 17, 24, 26, 27, 29, 30, 31, 32, 33, 34, 35, 36, 37, 38, 39, 40, 43, 44, 45, 46, 47, 48, 49, 50, 51, 52, 53, 56, 57, 59, 60, 61, 63, 64, 65, 66, 67, 70, 71, 72, 73, 74, 75, 77, 78, 79, 80, 81, 82, 83, 84, 85, 87
communication, 86
competition, 15
compliance, 32, 44, 48, 68, 75, 83
confidence, 39
congestion, 45
Congress, 3, 6, 7, 56, 57, 85
consensus, 23
conspiracy, 51
consultants, 4
consumers, 2, 11, 23, 39, 40, 53, 81
consumption, 20, 23
convergence, 14
corporations, 81
costs, 12, 23, 80
counsel, 68
Court of Appeals, 85
credit, 81
crude oil, vii, 1, 2, 3, 5, 9, 10, 12, 14, 19, 20, 21, 22, 24, 26, 27, 29, 30, 37, 45, 59, 81
CTA, ix, 12
currency, 21, 79
customers, 3, 33, 53, 66, 67, 84

D

database, 3, 45
decisions, 24, 56, 70, 86
defendants, 50, 51, 84
definition, 15, 79, 80
deflation, 59
deflator, 59
delivery, 2, 6, 9, 10, 12, 13, 14, 23, 37, 45, 46, 47, 50, 63, 64, 79, 80, 81, 86
demand, 1, 2, 5, 6, 9, 10, 12, 19, 20, 21, 23, 24, 27, 48, 55, 60, 81, 85
Department of Agriculture, 50
Department of Energy, 4, 59
Department of Justice, ix, 4, 75
derivatives, vii, 1, 2, 3, 5, 6, 7, 10, 11, 12, 15, 16, 24, 26, 27, 33, 34, 35, 38, 39, 40, 50, 55, 56, 57, 59, 60, 64, 79, 81, 82, 83, 85
detection, vii, 3
deviation, 25, 60, 82
directives, 66
dispersion, 82
division, 49, 66
draft, 8, 80

E

economic growth, 20
electricity, 39, 48
emission, 38
employees, 46
energy, vii, 1, 2, 3, 5, 6, 7, 9, 10, 11, 12, 14, 15, 19, 20, 21, 23, 24, 26, 27, 31, 34, 35, 36, 37, 38, 39, 43, 44, 46, 48, 49, 50, 51, 55, 56, 57, 59, 60, 61, 67, 71, 79, 80, 81, 83, 85
Energy Information Administration (EIA), ix, 4, 14, 20, 21, 22, 23, 24, 46, 59, 60, 61
energy markets, 2, 3, 4, 27, 34, 35, 46, 55, 60
Energy Policy Act of 2005, 49
energy supply, 20
Enron, 50, 73
environment, 53
equity, 66

Index

ethanol, 23
evidence, 33, 37, 47, 51, 67, 68
evolution, 55
exchange markets, 3, 50, 56
exchange-based trading, 43
execution, 81, 85
exposure, 34, 66
expulsion, 70
extraction, 22

F

failure, 70
Federal Register, 33, 40, 61
Federal Reserve, 39, 49, 60, 81
Federal Reserve Bank, 49
finance, 60
financial institutions, 84
financial markets, 5, 9, 10, 11, 19, 79
Financial Services Authority, 51, 63
financial soundness, 16
firms, 15, 27, 33, 44, 46, 50, 66, 80
fluctuations, 9, 11, 85
focusing, 44
forecasting, 11, 12
fraud, 3, 6, 15, 36, 40, 43, 47, 52, 56, 86
fuel, 12
funds, 1, 2, 4, 6, 12, 15, 25, 33, 34, 35, 39, 53, 60, 80
futures, iv, vii, 1, 2, 3, 5, 6, 7, 8, 10, 11, 12, 13, 14, 15, 16, 19, 20, 23, 24, 25, 26, 27, 30, 31, 32, 33, 34, 35, 40, 43, 44, 45, 46, 48, 49, 50, 51, 52, 55, 57, 59, 60, 61, 63, 64, 65, 67, 69, 70, 73, 79, 80, 81, 83, 85, 86, 87
futures markets, vii, 1, 3, 5, 8, 11, 12, 14, 15, 19, 20, 23, 26, 27, 32, 34, 40, 43, 44, 49, 55, 57, 59, 80, 83, 85

G

gasoline, vii, ix, 1, 2, 3, 5, 9, 10, 11, 12, 19, 22, 23, 24, 26, 27, 28, 29, 30, 45, 59, 80, 81
global demand, 5, 19, 21, 55

global economy, 20
goals, 2, 8, 32, 52, 53, 57
gold, 28
government, 4, 22, 52, 60, 75
GPRA, 87
grand jury, 68
gross domestic product, 59
groups, 3
growth, 1, 3, 20, 21, 22, 23, 27, 37, 52, 53, 55, 56
guidance, 83
guilty, 51

H

heating, vii, 1, 2, 3, 5, 9, 10, 12, 19, 22, 24, 26, 27, 30, 45, 59, 81
heating oil, vii, 1, 2, 3, 5, 9, 10, 12, 19, 22, 24, 26, 27, 30, 45, 59, 81
hedge funds, 1, 2, 4, 6, 12, 15, 25, 35, 39, 60
hedging, 34, 40, 61, 85
Homeland Security, 27, 60, 82
Hub, Henry, 9
hurricanes, 13

I

id, 34
India, 20
indices, 34
Indonesia, 82
industry, 15, 16, 20, 23, 35, 37, 43, 46, 48, 56, 60, 61, 65
inelastic, 9, 10
inflation, 5, 59, 80
initiation, 33, 79
innovation, 6, 15
insecurity, 20
instability, 5, 19, 20, 24, 55
institutions, 84
instruments, 15, 39
integrity, 3, 15, 31, 32, 33, 40, 53, 55
IntercontinentalExchange (ICE), ix, 28, 36, 37, 38, 45, 60, 63, 64, 83, 84

intermediaries, 16, 33
International Monetary Fund, 84
interview, 60
investment, 2, 4, 11, 12, 19, 22, 27, 35, 39, 60, 79, 80
investment bank, 4, 27, 35, 39, 60
investors, 15, 34, 55
Iran, 22, 82
Iraq, 22, 82
isolation, 24

J

Japan, 20
joint demand, 25
jurisdiction, 3, 15, 48, 81

K

Katrina, 22
Kuwait, 82

L

laws, 3, 39, 52, 84
lead, 21, 22, 23, 27, 35, 48
leadership, 15
legality, 15
legislation, 15
Libya, 82
liquidity, 11, 27
litigation, 72
location, 10, 63, 64
London, 63
Louisiana, 9

M

management, 7, 33, 48, 52
manipulation, 2, 3, 6, 7, 15, 16, 32, 36, 40, 43, 44, 45, 46, 47, 49, 50, 51, 52, 53, 56, 65, 67, 72, 73, 74, 75, 85

market(s), vii, 1, 2, 3, 5, 6, 7, 8, 9, 10, 11, 12, 14, 15, 16, 19, 20, 21, 22, 23, 24, 25, 26, 27, 30, 31, 32, 33, 34, 35, 36, 37, 38, 39, 40, 41, 43, 44, 45, 46, 47, 48, 50, 51, 52, 53, 55, 56, 57, 59, 60, 61, 63, 64, 65, 66, 69, 70, 71, 72, 79, 80, 81, 83, 84, 85, 86, 87
market position, 45, 47, 48, 71
market prices, 11, 14, 31, 35, 36, 86
maximum price, 66
measurement, 11
measures, 2, 7, 8, 11, 25, 44, 47, 52, 53, 56, 57, 82
membership, 60, 70
mergers, 51
metals, 28, 34, 38, 59, 66
methyl tertiary, 23
Mexico, 21
Missouri, 82
money, 11, 12, 19, 25, 26, 27, 29, 30, 33, 80
Moody's, 66
morning, 44

N

nation, 51
natural disasters, 20
natural gas, vii, 2, 3, 4, 5, 9, 10, 13, 19, 24, 25, 26, 27, 30, 37, 39, 45, 48, 50, 59, 75, 81, 84
negotiation, 39
network, 44
New York, ix, 1, 3, 4, 10, 16, 44, 46, 49, 51, 59, 61, 63, 65, 66
New York Mercantile Exchange, Inc. (NYMEX), ix, 1, 2, 3, 6, 7, 10, 13, 14, 16, 25, 27, 28, 32, 37, 39, 43, 44, 47, 50, 59, 60, 61, 65, 66, 67, 68, 69, 70, 74, 80, 83, 84, 87
Nigeria, 22, 82
North America, 51, 70, 72, 87
Norway, 21

O

obligation, 6, 10, 63, 79

Index

off-exchange trading, 15, 34, 43, 56
Office of Management and Budget (OMB), ix, 7, 44, 51, 52, 53, 61, 87
oil, vii, 1, 2, 3, 4, 5, 6, 9, 10, 12, 13, 14, 19, 20, 21, 22, 23, 24, 26, 27, 29, 30, 35, 37, 39, 45, 48, 50, 55, 59, 60, 66, 80, 81
oil production, 5, 21
Oklahoma, 9
operator, ix
organization(s), ix, 15, 16, 24, 33, 53, 60, 67, 81, 82
Organization of the Petroleum Exporting Countries (OPEC), ix, 20, 21, 22, 82
oversight, vii, 1, 3, 4, 6, 15, 16, 31, 32, 35, 37, 39, 43, 44, 48, 56, 57, 61, 81, 83
over-the-counter (OTC), ix, 1, 2, 3, 6, 10, 11, 15, 27, 31, 34, 36, 38, 39, 40, 55, 56, 59, 61, 63, 64, 66, 78, 83

P

partnerships, 81
penalty(ies), 16, 50, 71, 72, 73, 74, 75
pension, 34
performance, 2, 7, 8, 10, 23, 44, 45, 52, 53, 56, 57
permit, 15, 34
petroleum, ix, 20, 50, 78
petroleum products, 20, 23
pipeline hub, 9
planning, 11, 12, 52
police, 67
political instability, 5, 19, 20, 55
pools, 84
power, 22
pressure, 2, 6, 14, 23, 27, 30
prevention, 16, 32
price changes, 11, 25, 55, 82
price deflator, 59
prices, vii, 1, 2, 3, 5, 9, 11, 12, 14, 15, 19, 20, 21, 22, 23, 24, 25, 26, 27, 30, 31, 34, 35, 37, 39, 40, 45, 48, 50, 55, 59, 60, 64, 65, 71, 74, 79, 80, 82, 86
private investment, 79
producers, 9, 10, 11, 12, 24, 60

production, 1, 5, 10, 20, 21, 22, 23
profit(s), 9, 11, 14, 26, 30, 80
program, 2, 7, 8, 16, 33, 34, 44, 45, 47, 51, 52, 53, 56, 57, 61, 65, 70
promote innovation, 15
propane, 51
protocols, 48, 84
proxy, 28, 53
public interest, 52, 55
public policy, 56
publishers, 48
purchasing power, 22

Q

Qatar, 82
qualifications, 16, 65

R

range, 3, 10, 56, 60, 71
ratings, 44
recall, 66
recordkeeping violations, 71
refiners, 4, 12, 60
refining, 5, 22
regional, 44
regulation(s), 3, 6, 15, 31, 32, 33, 52, 61, 63, 65, 67, 80, 81, 84, 85
regulators, 6, 39, 48, 49, 55, 61
Regulatory Commission, ix, 4, 60, 75
regulatory oversight, 31, 81
relationship(s), 3, 6, 11, 45, 48, 51, 56, 59, 65
relevance, 6, 8, 34, 35, 57
resolution, 70
resources, 7, 48, 52, 56, 60
restitution, 16
retail, 3, 12, 56, 81, 84
returns, 55
rice, 51, 59, 64
risk, 10, 11, 15, 22, 39, 50, 64, 71, 79, 80, 81
Russia, 21, 22

S

sabotage, 20
sales, 33, 48, 71, 74
sanctions, 68, 70
Saudi Arabia, 21, 82
science, 46
scores, 52
search, 55
Secretary of the Treasury, 81
securities, 79, 80
Securities and Exchange Commission (SEC), ix, 4, 39, 77, 78, 81, 87
self-regulation, 15
Senate, 27, 60, 82, 85
September 11, 20
series, 52, 64
settlements, 70
shares, 49
sharing, 48, 51
Shell, 70, 72
shortage, 24
sites, 46
spare capacity, 20, 21
speculation, 6, 27, 85
spot market, 9, 10, 14
staffing, 16
standard deviation, 25, 60, 82
standards, 4, 23, 47
storage, 20, 23, 24, 70, 75
strategic planning, 52
strength, 65
subpoena, 49, 84
summer, 9
supervisors, 47, 48
supply, 1, 2, 5, 6, 9, 10, 12, 13, 19, 20, 21, 22, 23, 24, 25, 27, 45, 48, 55, 60, 81, 85
supply disruption, 13, 19, 21, 22, 23, 55
surging, 30
surplus, 21
surveillance, ix, 2, 3, 7, 8, 15, 16, 31, 32, 43, 44, 45, 46, 47, 48, 50, 56, 57, 61, 65, 66, 67, 69, 70
susceptibility, 32
systemic risk, 15

T

tanks, 12
targets, 52
technology, 24
telephone, 67, 86
terrorist, 20, 24
terrorist acts, 20, 24
Texas, 9
theory, 30
threshold(s), 36, 37
time, ix, 1, 5, 8, 10, 11, 13, 16, 23, 26, 27, 29, 49, 53, 55, 66, 68, 84
trade, 4, 6, 10, 11, 16, 27, 36, 37, 39, 47, 60, 61, 63, 65, 66, 67, 69, 70, 80, 85, 87
trading, vii, ix, 1, 2, 3, 6, 7, 12, 13, 15, 16, 19, 24, 25, 27, 28, 29, 30, 31, 32, 33, 34, 35, 36, 37, 38, 39, 40, 43, 44, 45, 46, 47, 49, 50, 55, 56, 57, 59, 60, 61, 64, 65, 66, 67, 68, 71, 72, 73, 74, 75, 79, 80, 81, 84
transactions, 10, 11, 13, 14, 15, 31, 32, 34, 36, 37, 38, 39, 40, 48, 63, 71, 83, 84, 85
transmission, 39, 48, 86
transparency, 6, 8, 12, 15, 31, 34, 40
transportation, 11, 20, 75
trend, 1, 23, 26
trustworthiness, 39

U

U.S. economy, 2
uncertainty, 11, 12, 20
United Arab Emirates, 82
United Kingdom, 21, 51
United States, 3, 9, 15, 20, 21, 23, 50, 51, 63
users, vii, 3, 4, 6, 10, 15, 33, 35, 52, 60

V

values, 10
variability, 25
Venezuela, 22, 82
volatility, 1, 11, 19, 23, 24, 25, 26, 27, 30, 34, 59, 82

W

Washington, 4, 16, 77, 78, 81, 82, 84, 85, 86, 87

wells, 10
winter, 13
Wisconsin, 82
wrongdoing, 53, 67